▶ **Software, Animation and the Moving Image**

DOI: 10.1057/9781137448859.0001

Other Palgrave Pivot titles

Mo Jongryn (editor): MIKTA, Middle Powers, and New Dynamics of Global Governance: The G20's Evolving Agenda

Holly Jarman: The Politics of Trade and Tobacco Control

Cruz Medina: Reclaiming Poch@ Pop: Examining the Rhetoric of Cultural Deficiency

David McCann: From Protest to Pragmatism: The Unionist Government and North-South Relations from 1959–72

Thijl Sunier and Nico Landman: Transnational Turkish Islam: Shifting Geographies of Religious Activism and Community Building in Turkey and Europe

Daria J. Kuss and Mark D. Griffiths: Internet Addiction in Psychotherapy

Elisa Giacosa: Innovation in Luxury Fashion Family Business: Processes and Products Innovation as a Means of Growth

Domagoj Hruška: Radical Decision Making: Leading Strategic Change in Complex Organizations

Bjørn Møller: Refugees, Prisoners and Camps: A Functional Analysis of the Phenomenon of Encampment

David Ralph: Work, Family and Commuting in Europe: The Lives of Euro-commuters

Emily F. Henderson: Gender Pedagogy: Teaching, Learning and Tracing Gender in Higher Education

Mihail Evans: The Singular Politics of Derrida and Baudrillard

Bryan Fanning and Andreas Hess: Sociology in Ireland: A Short History

Tom Watson (editor): Latin American and Caribbean Perspectives on the Development of Public Relations: Other Voices

Anshu Saxena Arora and Sabine Bacouël-Jentjens (editors): Advertising Confluence: Transitioning the World of Marketing Communications into Social Movements

Bruno Grancelli: The Architecture of Russian Markets: Organizational Responses to Institutional Change

Michael A. Smith, Kevin Anderson, Chapman Rackaway, and Alexis Gatson: State Voting Laws in America: Voting Fraud, or Fraudulent Voters?

Nicole Lindstrom: The Politics of Europeanization and Post-Socialist Transformations

Madhvi Gupta and Pushkar: Democracy, Civil Society, and Health in India

George Pattison: Paul Tillich's Philosophical Theology: A Fifty-Year Reappraisal

Alistair Cole and Ian Stafford: Devolution and Governance: Wales between Capacity and Constraint

Kevin Dixon and Tom Gibbons: The Impact of the 2012 Olympic and Paralympic Games: Diminishing Contrasts, Increasing Varieties

Felicity Kelliher and Leana Reinl: Green Innovation and Future Technology: Engaging Regional SMEs in the Green Economy

Brian M. Mazanec and Bradley A. Thayer: Deterring Cyber Warfare: Bolstering Strategic Stability in Cyberspace

DOI: 10.1057/9781137448859.0001

palgrave▶pivot

Software, Animation and the Moving Image: What's in the Box?

Aylish Wood
University of Kent, UK

palgrave
macmillan

DOI: 10.1057/9781137448859.0001

First published 2015 by
PALGRAVE MACMILLAN

Palgrave Macmillan in the UK is an imprint of Macmillan Publishers Limited,
registered in England, company number 785998, of Houndmills, Basingstoke,
Hampshire RG21 6XS.

Palgrave Macmillan in the US is a division of St Martin's Press LLC,
175 Fifth Avenue, New York, NY 10010.

Palgrave Macmillan is the global academic imprint of the above companies
and has companies and representatives throughout the world.

Palgrave® and Macmillan® are registered trademarks in the United States,
the United Kingdom, Europe and other countries.

ISBN: 978-1-137-44886-6 EPUB
ISBN: 978-1-137-44885-9 PDF
ISBN: 978-1-137-44884-2 Hardback

A catalogue record for this book is available from the British Library.

A catalog record for this book is available from the Library of Congress.

www.palgrave.com/pivot

DOI: 10.1057/9781137448859

Contents

DOI: 10.1057/9781137448859.0001

List of Figures

DOI: 10.1057/9781137448859.0002

Acknowledgements

I would like to thank all the people I interviewed. They gave me their time and shared their enthusiasm about what they do when creating animations. I also acknowledge funding from the Arts and Humanities Research Council, whose support made those interviews and my wider research possible.

▶

palgrave▶**pivot**

www.palgrave.com/pivot

Introduction

Wood, Aylish. *Software, Animation and the Moving Image: What's in the Box?* Basingstoke: Palgrave Macmillan, 2015. DOI: 10.1057/9781137448859.0004.

▶

What do we already know about computer-generated animation? When *Frozen* (USA, 2013) or *The Lego Movie* (Australia, 2014) gain widespread acclaim, and publicity surrounding their success sparks various debates (about princesses in Disney or the nostalgia of Lego), the look of computer generated figures and storytelling practices have a wide reach. Such high profile features are only the tip of what's out there, as computer animation is commonly behind visual effects in live-action films, forms the basis of all kinds of computer games, is widely used in making adverts and data visualizations, and found on many websites too. What we know about all these computer-generated animations is often based on what we see on the screen. Frequently this tells us about an increasing capacity to simulate or depict physical reality, whether as extraordinarily photoreal imagery in films such as *Dawn of the Planet of the Apes* (USA, 2014), amusing adverts like *The Pony* (UK, 2013) for Three (featuring a moon dancing Shetland pony), or the spread of touch screen smart phone and tablet games including *Angry Birds* (Finland, 2009), *BADLAND* (Finland, 2013) or *Monument Valley* (UK, 2014). Through television news reports and documentary based programs, animation's facility to inform is visible as well. From this diverse range of increasingly pervasive animation, we can see and hear the ways in which computer animation software is a highly pliable technology whose capacity to become embedded in existing forms of storytelling, information sharing and marketing is fully exploited.

When thinking about the software behind the creation of all of these images, things become more opaque. There is a tendency to look at software not so much in terms of what it is, but what it does. Amelia, IPSoft's 'cognitive agent', described as their first to understand like a human, illustrates this very nicely. As IPSoft put it: 'Amelia, our cognitive knowledge worker, interfaces on human terms. She is a virtual agent who understands what people ask – even what they *feel* – when they call for service' (IPSoft, 2014). As an image on a screen, Amelia is a virtual model with which many are familiar, an avatar clearly identifiable as a young, white woman with blond hair. Behind this façade exists a complex A.I., a set of algorithms not mimicking the human brain, but rather 'recreating how our brains comprehend'. While there is a lot to unpack in the idea that algorithms might recreate how a brain comprehends, Amelia is mentioned here to only pose an equivalent question about computer animation software. Commentators often pay attention to how well images made using computer-generated animation mimic or simulate physical reality. But

DOI: 10.1057/9781137448859.0004

computer animation software recreates physical reality through a process that successfully draws together the creative skills of animators and the complexities of the software they are using. Underlying this process are numerous assumptions and abstractions, and multiple sets of influences, both human and technological. Getting to know the complexities of software is central to *Software, Animation and the Moving Image*, and is used to explain how animators and software fold together in the action of making moving images. Knowing software means not simply thinking in terms of its code and what it produces, but tracking its appearance in paratexts such as training materials, publicity and marketing strategies, or commentaries on particular examples of its use, through interviews with its users, as well as images created when using the software.

Before coming onto questions of software, I want to first clarify what I mean by computer-generated animation, which when used without further elaboration is actually quite vague. In the same way animation comes in all kinds of shapes and forms, so does computer-generated animation and the software used in its production. Even though many might assume it refers simply to 3-D animation, computer-generated animation is an umbrella term for moving images created using a diverse set of packages sharing the facility of relying on algorithmic processes as part of producing artwork. These can be packages developed for children or adults, 2-D or 3-D animation used on computers (Toon Boom, AfterEffects, Flash), the web (Go Animate and Muvizu) or iPad (Animation Studio and Animation Desk), aimed at professional (Maya, 3DS Max, Blender) or non-professional animators.[1] Adding another layer of ambiguity, with the release of *Up* (USA, 2009) and *Toy Story 3* (USA, 2010), in only the last few years 3-D has also come to mean animations created for 3-D projection, even if they are a stop-motion animation such as *Coraline* (USA, 2009) or *ParaNorman* (USA, 2012). In *Software, Animation and the Moving Image*, the computer-generated animations I primarily discuss are ones in which 3-D figures inhabit a three-dimensional environment. 3-D animation is currently a dominant technique in animation, used for visual effects, data visualizations, games and adverts as well as animations. As such, it a good place to start asking questions about software, animations and moving images. The specific software studied is Autodesk Maya, and while many of the observations about Maya and working in 3-D space are relevant to other 3-D software, they are not generally applicable to moving images created using 2-D animation software.[2]

DOI: 10.1057/9781137448859.0004

Although *Software, Animation and the Moving Image* takes the 3-D animation software Autodesk Maya as its case study, there are many other 3-D animation packages available. Competitors include Cinema 4D, Houdini, Autodesk 3DS Max, Z-Brush and XSI: SoftImage.[3] I should say up front, I am not seeking to argue that Maya is distinct from or better than other 3-D software. But, as an example of complex multifunctional software, Maya is an ideal study for several reasons. Maya retains a pre-eminent position in the creative industries, making it a good candidate as a case study for exploring a wider set of ideas about 3-D animation. First released in 1998 as Alias|Wavefront Maya, the software quickly won plaudits within the film industry through its use in creating the Oscar winning effects in *The Matrix* (USA, 1999). A strong marketing campaign, and a strategy of targeting educational institutions with license fee reductions, which continues today through free educational access, ensured the software gained traction in both the professional and training-to-be-professional sectors. In 2003, Alias|Wavefront won an Oscar for the development of the software. Taken over by Autodesk in 2006, Maya continued to undergo annual updates and, as well as selling as a stand alone product, it is now packaged as part of the Autodesk Entertainment Creation suite alongside 3DS Max, Motion Builder and Mudbox. Maya's strong presence across visual effects, animation, adverts, TV and data visualizations is sustained by its versatility and extensibility. The package includes software for modelling, animating, rigging, lighting and rendering.[4] It is also open, which means production houses and individuals customize Maya by writing scripts to automate repetitive tasks or enhance a feature to match the needs of individual projects. Although now more vulnerable to competition, Maya remains widely used and taught and has associated with it an extensive range of on-line tutorials and numerous forums through which people give advice, solve problems, and share scripts.[5] With such a rich expanse of material available, Maya is a long-standing and widely used software well suited for exploring ideas circulating about 3-D animation software.

Approaching software

Writing about digital humanities Rieder and Röhle comment: 'digital technology is set to change the way scholars work with their material, how they see it, and interact with it' (Rieder and Röhle, 2012, p. 70). This

DOI: 10.1057/9781137448859.0004

remark also rings true for animators working with software; it changes the ways in which animators (modellers, riggers, animators, technical directors) work with the entities they create, how they see it and how they interact with it. To really explore what this means, some account of software is necessary. But how does one go about giving an account of a software package? What does it mean to study software, something that is only visible indirectly?

Starting from software may seem a strange, even estranging place to begin, especially for people used to looking at what software produces, the moving images of live-action visual effects, animations, games and data visualizations. Staging an encounter with software might at first sight be an unexpected notion, as a very common perception of software is of algorithms, two-dimensional lines of code scrolling on a screen or written on paper, sets of instructions telling a computer to act in particular ways. These instructions can include the operating systems of computers, word processing packages or the many kinds of software used to create different kinds of animations. The description of software simply as a set of instructions leaves aside much of the complexity of software. As a counterbalance, software studies sets about the task of complicating software, seeing it not simply as a set of instructions, but as an object inflected with social and cultural influences. In *Behind the Blip*, a key early work in software studies, Matthew Fuller argued that the invisible walls of software could be broken through to reveal another universe, one unclassifiable but seething with life (Fuller, 2003, p. 48). As he put it, the difficulty is: 'Software lacks the easy evidence of time, of human habitation, of the connotations of familial, industrial, or office life embedded in the structure of a building' (ibid.). Since these early moments in software studies, scholars have found numerous ways of conceiving of software and the ways in which it reveals the evidence of time and human habitation.

Adrian Mackenzie's study of software as a cultural process is an important touchstone for my project. As a means of understanding code as practice and material, his analyses took in many paratexts surrounding software, so negating the tendency to perceive of software simply as an application. Mackenzie comments: 'Software ... looks increasingly like a neighbourhood rather than an intangible, abstract formalism' (2006, p. 3). By seeing software as a neighbourhood, it becomes visible as a multidimensional and shifting object. Different practices of production, consumption, use and circulation give rise to a code 'permeated by all the

forms of contestation, feelings, identification, intensity, contextualiza-
tions and decontextualizations, signification, power relations, imaginings
and embodiments that comprise any cultural object' (ibid., p. 5). Wendy
Chun stakes a similar claim when she says: 'software is best drawn in
terms of position in a field of relations. Software is "in media res – in
the middle of things"' (Chun, 2011, p. 176). Looking at software in the
middle of things, as encountered through a neighbourhood of relations
is my approach in *Software, Animation and the Moving Image*, widening
the appearance of software in our discussion of computer animation,
and re-thinking the images through software. A variety of materials
have been used to explore Maya, including video tutorials created by
both authorized trainers (Escape Studio and CG Society) and numerous
on-line examples uploaded to YouTube and created by members of the
wider Maya community of users.[6] Manuals have also been consulted,
though on-line videos are a more widely used source of information by
users. These materials, both official and informal, form the backdrop of
my discussion of Maya. Between them, paratexts provide a discourse
around the software, through which permeate contestations, power relations and embodiments, expressions of the production culture of both
the producers of Maya and also the users of the software.

Claiming cultural practices of media production as an important
site of analysis, John Caldwell describes the media industry as having
workaday forms of critical and cultural analysis that provide insights
for media scholars (Caldwell, 2008). Through their production culture,
production communities generate cultural expressions, 'involving all of
the symbolic processes and collective practices that other cultures use: to
gain and reinforce identity, to forge consensus and order, to perpetuate
themselves and their interests, and to interpret the media *as* audience
members' (ibid., p. 2). At times, these reflections reveal complex thinking about the processes involved. At other times, the reflections are
more straightforward descriptions of innovative technologies or solu-
tions to problems. In either case, these disclosures form texts that can
be analysed; they are narratives about the elements involved in a film's
production. The same claims are true of cultural practices surrounding
Autodesk Maya, as user communities and producers of software contribute to the various threads comprising the wider cultural perceptions
of the software. Production culture scholars caution against taking any
statements at face value, to understand them instead as managed, parts
of the spin that more widely inform industrial disclosures:

DOI: 10.1057/9781137448859.0004

> One must be mindful that all texts, whether found in an archive or one's own field notes, are constructions, versions on the real that may serve different roles in a production studies project, from corporate branding and spin, to the personal reflections of an outsider looking in. (Mayer et al., 2009, p.10)

Taken in this light, production culture disclosures are narratives open to cultural analysis, equally revealing facts about what happened during production, the local and global politics surrounding that production, and also embedded power structures. Autodesk Maya, rooted in the visual effects, games and animation industries, is part of the local and global politics of changing work practices in these industries. Within this context, there are narratives about poor wages, precarious employment, exploitative practices, reduced status as a creative contributor in an automated pipeline, as well as narratives of great innovation and creative insight. The focus of *Software, Animation and the Moving Image* remains at a local scale, in the sense that it pays attention to what users say they do, and are encouraged to do, through various cultural disclosures. In the first chapter, the particular theme pursued is creative practices in auto- mated contexts, and the ways this plays out across the various cultural disclosures found in both the history of software and individual users of contemporary software.

Even though Maya can to an extent be known through various para- texts, it also needs to be understood as software. As Adrian Mackenzie remarks: 'what software does and how it performs, circulates, changes and solidifies cannot be understood apart from its constitution through and through as code' (2006, p. 3). Talking about code and software, David Berry notes: 'code and software are two sides of the same coin, code is the static textual form of software, and software is the processual operating form' (2011, p. 32). In making sense of Maya as a processual operating form, I think about the software through toolsets on the user interface. Working with toolsets to find out more about software is also an approach developed by Lev Manovich. Manovich's *The Language of New Media* is often cited as an inaugural work in software studies, and through his ongoing interest in moving image software, including specific studies of Aftereffects and Photoshop, he continues to ask how software changes media, concluding that image-editing operations have 'not just one but two parents, each with their own set of DNA: media and cultural practices, on the one hand, and software development, on the other' (Manovich, 2011, p. 11). Drawing out the idea of 'deep remix- ability,' he remarks software like Maya offer users a new experience of

DOI: 10.1057/9781137448859.0004

image-making, one that 'cannot be simply reduced to a sum of the working methods already available in separate fields' (ibid., p. 285).

Toolsets and their operations are also the concern of *Software, Animation and the Moving Image*, but my encounters with them is staged in a different way, through practices and contexts from which they accumulate meaning. This perspective situates the formal properties of code and operations as being 'in-use', concentrating on the routes through which a seemingly abstract entity based on code becomes part of a meshwork of meaningful structures. To do this, Maya's user-interface is viewed not just as an array of toolsets, but also as a visual organization attaching specific frames and patterns to the abstract space and time of the algorithmic processes. Making sense of the user-interface in this way draws on Noah Wardrip-Fruin's account of organizational logic in his work *Expressive Processing* (2009). He describes how a player's encounter with a game involves the interplay of: 'data, process, surface, interaction, author, and audience' (Wardrip-Fruin, 2009, p. 13). By giving an account of the virtually projected shapes seen in the viewport, and the schematic depictions of software operations that co-exist with the viewport, Maya's user interface is analysed as a relay of interactions between a surface of toolset menus and the deeper algorithmic structures where data is processed. This analysis provides the basis on which to further examine the history of software and the experiences of contemporary users. My discussion of the latter is based on interviews carried out with 20 Maya users (18 face to face, two via email), all of which inform this study. The interviewees work in a range of animation sectors: major fx houses (The Mill, Double Negative), small, medium and larger sized animation studios and organizations (Digimania, Aardman, and Institute of Creative Technology), games studios (Blizzard Entertainment and thatgamecompany), data visualization studios (GST, Inc. – NASA/Goddard Science Visualization Studio, Harvard Medical School, the Walter and Eliza Hall Institute of Medical Research, and PinPoint Visualization), as well as individual animators and also students enrolled on undergraduate programs.

Looking at the visual organization of Maya's user interface and working with interview materials goes some way to: 'show the stuff of software in some of the many ways that it exists, in which it is experienced and thought through' (Fuller, 2008, p. 1). In the first chapter, an account is also given of the history of software and computer animation. The story of how computer animation gained traction in the film industry is well known (Finch, 2013; Sito, 2013; Catmull, 2014), so I am not going to repeat it here in any detail. Instead, the history of 3-D software, more specifically its timing algorithms, is explored as a way of more fully understanding how a user and software fold together during the process of making things move and building models. Since the late 2000s, platform studies has also emerged as an area of study:

> Platform studies is an approach that looks at the relation of hardware and software as a system, from the electronics inside the console box to the peripheral controllers, and at how the affordances and constraints of a particular system invite as well as shape the development of creative works. (Jones and Thiruvathukal, 2012, p. 4)

Though not a feature of my study, the platforms on which Autodesk Maya was developed and is used do matter in a wider history of the software. For instance, following a merger between Alias (the early developer of Maya) and Silicon Graphics Inc. (SGI), the two worked to market the combination of Alias software and the new microprocessors developed by SGI. A platform studies inflected history could take into account the relationship between Maya's development and SGI's IRIX operating system, and the subsequent migration of Maya to Windows-DOS, Mac-OS, and also Linux. It might also want to take into account the introduction of multi-core processers and the ways in which Maya's developers optimized code to take advantage of the additional processor power and multithreaded tasking with the release of Autodesk Maya 2009 (Autodesk, 2008a). Nick Montfort and Ian Bogost comment: 'whatever the programmer takes for granted when developing, and whatever, from another side, the user is required to have working in order to use particular software, is the platform' (Montfort and Bogost, 2009, p. 2). This study continues to take the platform somewhat for granted, engaging instead with the software through the user interface. For my purposes, the user interface is the primary location through which the frames and patterns of Autodesk Maya become visible to its users.

Approaching moving images

By getting closer to software through analysis of paratexts and the visual organization of the user interface, as well as gaining insights from interviews with users of Autodesk Maya, two novel avenues for thinking

DOI: 10.1057/9781137448859.0004

about animation open out. Although computer-generated animation is already widely discussed, the complexity of software means that relatively little is written about the 'doing' of animation, which at times leads to a series of assumptions and preconceptions. Using interviews to illuminate computer-generated animation re-sets the balance. Giving access to animators' understanding of both computer-generated animation and how they creatively work with software reveals some of the story of how computer animators animate.

As explored in the second chapter, knowing more about software not only gives greater insights into how computer-animations are created, it also offers another dimension for thinking about computer-generated animated images. Even as ever more sophisticated imagery is created by users of software, with highly complex modelling, texturing and photoreal rendering acting as key drivers, the difference software makes is drawn attention to in distinctive ways. As Lisa Purse puts it: 'the "digital-ness" of the digital image has the potential to produce connotations of its own' (Purse, 2013, p. 14). The possibilities for seeing this digital-ness in computer-generated imagery depends on the extent to which it sits outside of conventional behaviours associated with the laws of physical reality, or what we might call a logic based on realism: 'Automated algorithms provide the spaces, objects and even "camera" with a set of behaviours consistent with the physics of the real world at the same time as they allow for a plasticity in such rules only possible in animation' (Gurevitch, 2012, p. 134). By taking a software-centred approach, *Software, Animation and the Moving Image* develops a distinct analytic viewpoint. Working in the digital space of software generates movements and proximities named here as digital contours. Informed by the digital contours of software protocols, and drawing on ideas in cinema and animation studies, I aim to bring greater definition to something that remains implicit in many discussions of computer-generated images: a more-than-representational space with digital origins and whose appearance on the screen adds another affective dimension to our experience of moving images.

Notes

1   Stop-motion animation packages are also increasingly available, from those aimed at children such as Zu3D and Hue Animation Studio, to those such as Dragon used for higher end production.

DOI: 10.1057/9781137448859.0004

2 Autodesk Maya is a 3-D animation software, which can be used to create images for both 2-D and 3-D projection. Further details about its history and toolsets are given throughout the discussion.

3 As of early 2014, Autodesk has withdrawn Softimage from the market. Softimage was very long-standing, pre-dating Maya, and its withdrawal has been greeted with affectionate dismay (Burns, 2014).

4 Casey Alt and David Ryan provide some background to the development of Autodesk Maya (Alt, 2002; Ryan, 2011).

5 Maya is an open software in the sense that users can write scripts to help with their modelling, animation, rigging or lighting. Some users share these scripts with the Maya community.

6 I am grateful to both Escape Studio and CG Society for giving me access to samples of their training materials.

DOI: 10.1057/9781137448859.0004

1
Getting to Know Software: A Study of Autodesk Maya

Abstract: *Drawing on software studies, Wood offers an original methodology for approaching software by analysing the user interface of the 3-D animation software Autodesk Maya in the context of its paratexts. This entails scrutinizing the operational logic of the software as it appears on the user interface, the frames and patterns associated with the surface and also the deeper structures of the software. In Getting to Know Software, Wood also looks back into Maya's history and gives an account of the context of Autodesk Maya's production and release, and a media archaeology of movement algorithms more generally. The discussion draws on interviews with users of software. This material provides many insights into what animators do as they create moving images using software.*

Wood, Aylish. *Software, Animation and the Moving Image: What's in the Box?* Basingstoke: Palgrave Macmillan, 2015. DOI: 10.1057/9781137448859.0005.

What does software look like? A first thought might be that software never looks anything at all, since it exists inside a computer. It can only be seen through some kind of mediation, such as lines of executing code visible on a graphical user interface, running down the screen Matrix-like in films or TV shows, or, as an imaginative interpretation. Google the word 'digital', select images, and you will likely find an array of similar imaginative interpretations popping up. These frequently feature zeros and ones, perhaps arranged like a tunnel running off into the distance or as an electronic aurora borealis with zeros and ones embedded in bright white strands of light. *Wreck-It Ralph* (USA, 2012), an animated feature about game characters, goes a little further when it briefly depicts a game's software. The dastardly King Candy reigns over the code of Sugar Rush, and, seeking to extend his control, goes into the space where the software is being executed (Figure 1.1).

Interestingly, the space is depicted as having no gravity. As King Candy twirls onto his back, its dimensions are uncertain, with spatial definition given only in a central mass of connecting strands that look rather like cable. Closer too, these strands connect semi-translucent 3-D boxes that flash on and off, presumably to indicate when they are active or not. The strands and boxes are reminiscent of flow diagrams. Such a depiction of software is effective in capturing a common understanding of software as something that carries data, works in multiple ways at the same time, and is almost transparent and rather abstract.

FIGURE 1.1   *Digital space in* Wreck-It Ralph

Note: Still from *Wreck-It Ralph* (2012) showing King Candy floating in the gravity-free and dimensionally indistinct space of software.

Source: Image ©Disney Studio.

DOI: 10.1057/9781137448859.0005

Finding out more about software is the work of this chapter. Without unpicking its code, my approach is to give an account of software through its user interface (UI), where its logic becomes legible and meaningful. As described in the introduction, software is far from neutral. Adrian Mackenzie makes an equivalent point when writing about programs and operating systems such as Java and Linux: 'The incorporation of prior knowledges, codings and practices within the sequence of operations, attaches specific frames and patterns to the relatively abstract space and time of algorithmic processes' (Mackenzie, 2006, p. 64). Noah Wardrip-Fruin and Ian Bogost too advance a similar insight by examining how frames and patterns can be discernible in a player's encounter with computer games. Noah Wardrip-Fruin reasons that expressive processing in digital media are legible examples of 'things that we need to understand about software in general' (Wardrip-Fruin, 2009, p. 5). Ian Bogost argues that the procedural systems of many different kinds of games generate behaviour grounded in rule-based models, mounting logic-based arguments influencing a player's decisions (Bogost, 2007). For users of software, the user interface is where frames and patterns with the potential to influence their decisions are most evident, where digital media are legible and software behaviour folds together with user behaviours.

In their inaugural work on platform studies, Nick Montfort and Ian Bogost comment:

> The interface, although an interesting layer, is what sits between the core of the program and the user; it is not the program itself. A chess program may have a text interface, a speech interface, or a graphical interface, but the rules of chess and the abilities of a simulated opponent are not part of the interface. (Montfort and Bogost, 2009, p. 145)

For anyone interested in a program or platform, an interface offers fewer opportunities for thinking about how those things work than do the things themselves. For this study, which is attentive to how software makes an appearance, and the ways in which users engage with software through that appearance, the UI is the most interesting layer. As a bundle of toolsets, it gives access to the processes of the multi-functional operations of software. In Autodesk Maya, the focus of this study, you can model, animate, light, rig, and render, and each of these suites of functionality have a vast array of toolsets. It would be reasonable ask whether or not a UI simply gives a user easy access to sets of tools.

DOI: 10.1057/9781137448859.0005

The answer is both yes and no. The main functional purpose of a UI is indeed to provide access to various toolsets, but at the same time, the UI reveals levels of operational logic associated with the programming of the software.

Noah Wardrip-Fruin, in his work on expressive processing, a software studies approach to computer games, introduces the framework of operational logic:

> Those working in commercial areas of digital media, such as computer games, construct systems that operationalize ideas of narrative structure, character behavior, linguistic interaction, and so on. Each of these is something that, in other domains, we are accustomed to scrutinizing closely, often seeking to understand something of their underlying logic. But in the area of software, in which the underlying logic exists in an explicit encoding that can be examined, this takes place very rarely. (Wardrip-Fruin, 2009, p. 320)

My first tactic in making Autodesk Maya less abstract is to tease out the operational logic that appears on its user interface. Scrutinizing the options available to users and their mode of presentation reveals the frames and patterns associated with the surface and also the deeper structures of the software. Through these, I consider how the interface puts into operation ideas around creative work in the context of an automated system. Excavating the operational logic of Maya's UI is only a starting point for making the software less abstract. As David Berry says: 'The challenge is to bring software back into visibility so that we can pay attention to both what it is (ontology), where it has come from (through media archaeology and genealogy) but also what it is doing (through a form of mechanology ...)' (Berry, 2011, p. 4). Knowing about Maya's operational logic tells us about what it is, and forms the focus of the first section of the following discussion. Gaining an understanding of where Maya comes from entails looking back into its history, both of the software specifically and computer animation software more widely. Since Maya is a large and multifunctional software, my focus is specifically on timing. By looking at the different drivers behind software development, how this history has become embedded in its operational logic, Maya's UI is thought through as a network of interlocking discourses. To establish this perspective, in section two, I assemble the context of Maya's production and its release, along with an archaeology of movement algorithms. Finally, to make what software does less of a series of abstract processes, I turn to the insights gained from interviews with users of software. This

DOI: 10.1057/9781137448859.0005

allows me to think through how and where a user and software interact with each other and to consider the ways in which agency is distributed between user and software.

Locating the operational logic of Autodesk Maya

Lev Manovich too is interested in the connection between a UI and program and comments that the principles of a program are projected onto its UI: 'The principles of contemporary computer programming are 'projected' to the UI level – shaping how users work with media via software applications practically, and how they understand this process cognitively' (Manovich, 2013, p. 219–220). The UI offers an array of toolsets to anyone working with software. In Maya, the toolsets can be accessed through various modes of presentation. For instance, when modelling, toolsets give access to the modelled entities as both shapes and also packages of data. Maya user Ben Thomas notes the different ways the UI allow him to engage with objects he has created: 'I like to see it with the interface [the viewport]. You can do things in hypergraph, where you are viewing lots of texts and names and numbers, but coming from the background that I do, I prefer to see it in front of me'. For the moment my focus will be on the viewport (the hypergraph is returned to later), where Thomas can see the objects with which he is working. The viewport is designed to give users an impression of direct interaction with the shapes they create using various modelling tools sets. In the viewport, as the image of a 'sack-in-space' shows, modelled shapes appear as volumetric objects projected in virtual space, and any inputs from the user almost instantaneously appear on the screen (Figure 1.2).

Whether used in its three-dimensional perspective or orthogonal view, within this framed space, a user's access to and control over the object is central. When working within a perspective view, a modeller can see their model from all sides, as well as zoom in and out to increase or decrease the detail. Referring to his experiences of modelling, Barry Sheridan of Digimania comments: 'I still like having something 3-D in front of you and being able to spin around, I still like all that stuff, I still like the immediacy of being able to manipulate something seemingly floating in front of you'. Sheridan describes here the accessibility of the viewport, where users can see their model virtually projected, and where their interactions generated using a set of manipulators are immediately

DOI: 10.1057/9781137448859.0005

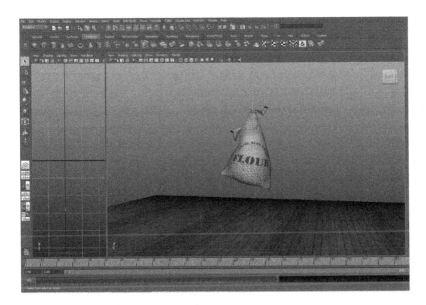

FIGURE 1.2   *Image projected in Autodesk Maya's viewport*

Note: The brown sack is a virtual projection in 3-D space, as seen in the perspective view in Autodesk Maya's viewport.

Source: Image ©Paul Hilton.

visible on the projected shapes. Models in the viewport are created using three different basic toolsets: polygons, NURBS, and subdivision surfaces. Which set the modeller works with depends on the kind of object they want to create, but usually it involves beginning with a series of primitive shapes to build the space filled by the model. Once the basic shape is in place, various other tool sets can be used to add more nuanced detail. For instance, Artisan is a feature within Maya were users access a sculpt tool to push, pull, and smooth areas of the model. As described in *The Art of Maya*: 'Painting with Artisan is an intuitive way to make selections, sculpt surfaces and paint values on selected surfaces or multiple surfaces' (Autodesk, 2008b, p. 52). The description of Artisan as intuitive echoes the observations of users for whom the viewport and various manipulators can be combined to give a sense of access and control. It gestures towards the familiarity of real-world space by projecting volumetric objects and allowing control over constructed spaces.

Such description and comments present the viewport as a place where artists are able to fully manipulate their objects. This emphasis fits not

DOI: 10.1057/9781137448859.0005

only with the functionality of the viewport, how it works, but doubles as a revelation of the logic expressed by the viewport's processes. Expressive processing for Noah Wardrip-Fruin too doubles up as it conveys both the ways in which authors and artists can use the expressive potential of computational processes (as they use the viewport and other tools), and also that these processes themselves express meaning through their design and histories. When using Maya, artists clearly exploit the expressive potential of the software in creating animations. But in addition, these computational processes express 'humanly meaningful elements and structures' (Wardrip-Fruin, 2009, p. 156). They are not just toolsets allowing an artist to do something; rather their arrangements as toolsets are meaningful as opposed to abstract. Such a doubling of meaning is pursued here through an analysis of the visual organizations of the software's UI. According to Wardrip-Fruin: '... the concept of expressive processing includes considering how the use of particular processes may express connection with a particular school of thought... revealing connections with communities of thought in software engineering, statistics or a wide range of other areas' (2009, p. 158–159). Following this logic, the manipulability of the objects in the viewport connects with the aim of making software accessible for artists, providing the impression of something in front of them. Layout artist Ben Thomas also describes a sense of immediacy, but this time in the context of dressing a set within a computer when using Maya:

> What I loved about layout is that it was set dressing within the computer. You've got these buildings and assets or models. I was working on a sequence that involved London, the Thames North bank. It was my job to arrange these assets in a manner that the supervisor wanted. We had one set and then we could place multiple cameras within the sequence, so you didn't have to create one layout per shot. You had basically a set, and then you could place your cameras within that set.

As well as describing the immediacy of set dressing within the computer, Thomas is also gesturing towards the efficiency of using a computer to generate a virtual model that can be modified for different cameras, as opposed to creating several layouts for multiple cameras.

As users of Maya describe accessible objects seeming to float, or a set layout configured within the viewport, each conveys an impression of immediacy even though both activities are highly mediated. Paying more attention to the processes of mediation, and how they generate this sense of immediacy brings the operational logic of the software into the open.

DOI: 10.1057/9781137448859.0005

Ian Bogost's notion of procedural rhetoric is helpful for thinking in closer detail about the ways software mediates: 'Procedural rhetoric, then, is a practice of using processes persuasively. More specifically, procedural rhetoric is the practice of persuading through processes in general and computational processes in particular' (Bogost, 2007, p. 3). Understood to be a technique for making arguments with computational systems, the procedural rhetoric of Maya's viewport is to persuade users of a seemingly direct interaction with objects and entities with which they are working. The mediations of the technology sit in the background as the artistry of the modeller or animator is foregrounded. To complicate my claim, the logic of this rhetoric is not, however, the only persuasive process in play on Maya's UI. It sits in direct contrast to another put forward through the dialog boxes sitting alongside, and frequently used in conjunction with the viewport. Figure 1.3 shows the viewport next

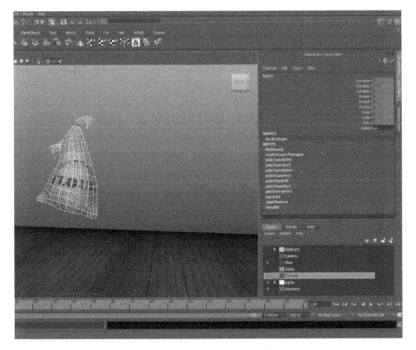

FIGURE 1.3   *Autodesk Maya's viewport with channel box*

Note: The shape of a sack projected in the viewport (left) alongside a dropdown dialog box that configures the image as data (right).

Source: Image ©Paul Hilton.

DOI: 10.1057/9781137448859.0005

to the channel box and layer editor (on the left), with the brown sack depicted as a shape as well as numbers and associated instructions. The presence of both persuasive processes creates a tension that establishes the operational logic of the software: creative work in the context of an automated system.

The dialog boxes, and their persuasive processes foregrounding the mediations of the software, are again explored as visual organizations on the UI. In addition to the viewport, which virtually projects the shape of any object, drop down windows augment immediate visual information about models, animation, rigging, lighting, shading or rendering. These drop down menus and dialog boxes demonstrate how Maya's programming deals with objects and movements as a series of datasets, rather than as shapes moving in space and time. Writing in 1998 about the UI design of Maya, George Fitzmaurice and Bill Buxton commented on their aim to expose the 'deep structure' of the software. Though aware that artists may not really want to know much about Maya's deep structure: 'providing access and acquiring such under-standing is often necessary for users to achieve their goals. Thus, we must find ways of exposing the deep structure to the user in ways that are compatible, intuitive and efficient' (Fitzmaurice and Buxton, 1998, p. 64). Consequently, the primitive shapes visible and accessible in the viewport sit alongside dialog boxes that could include the channel box/ attribute editor, the outliner, hypergraph, graph editor and the script editor. Dialog boxes depict schema showing relationships between packets of data constituting a scene, a contrast to the viewport where relationships are drawn between co-ordinates along lines of geometry, in other words as shapes. These schemas are logical configurations bringing deep structures closer to the surfaces. Layout artist Ben Thomas, who above expresses a preference for working in the viewport, also shared the following thoughts about the schematic aspects of the UI. His answer is framed in relation to a question about whether he thought the Maya's UI had a specific language:

> I think in terms of the language, I suppose it does in a way. In the sense of the way you go through a process, the way it guides you to do something. To a certain extent you are in control, but maybe it does dictate the way in which you do things to a certain object. And then, obviously, the more time you spend in there, you kind of understand. If you're modelling, that language is mirrored in rigging in the way that it needs to have something done to make it to work. I guess it's to do with the algorithms behind it, so the order in

DOI: 10.1057/9781137448859.0005

which you need to constrain an object to another object, and what you need to select first to do something.

Thomas's remarks reveal several features running counter to the immediacy of accessing objects via a viewport. His comments about constraints gesture towards ordering and hierarchies. He also talks about how the processes of the UI guide users to work in specific ways, and that enough of an understanding of how this works accumulates with more experience, a process linked to a growing awareness of the underlying algorithms of the software. The idea that processes of a UI can guide the ways a user approaches and thinks through a task will be developed more fully later in relation to the history of computer anima-tion software and also via interviews with software users. It is a fruitful way of gaining a more precise insight into how software mediates as a kind of entanglement. Before moving into that idea, I expand further on the ways in which software influences via its procedural rhetoric.

Linking the persuasiveness of procedural rhetoric to programming, Bogost says: 'its arguments are made not through the construction of words or images, but through the rules of behaviour, the construction of dynamic models. In computation, those rules are authored in code, through the practice of programming' (Bogost, 2007, p. 29). Unlike the situation encountered in a viewport, where the immediacy of an object is central, the mediation of software becomes much more explicit through the procedural rhetoric of dialog boxes and their various drop down menus. Where the viewport rhetoric seems to privilege working with and directly manipulating the model, dialog boxes promote an expanded arena in which the artist can work, albeit through a process more reliant on degrees of technical know-how. In *Persuasive Games*, Ian Bogost uses the phrase 'possibility space', which he again developed in relation to computer games: 'This is really what we do when we *play* videogames: we explore the possibility space its rules afford by manipulating the game's controls' (ibid., p. 43). Katie Salen and Eric Zimmerman defined play as: 'the free space of movement within a more rigid structure' (p. 304), and Bogost builds on their idea to argue that the processes of a software themselves create a possibility space. Such thinking is also relevant to users working with Maya's UI, the possibility space of the viewport sits alongside that of the dialog boxes. Within these two possibility spaces, two configurations of space co-exist: space as co-ordinates and shapes or space as packets of data. Figure 1.4 shows the now familiar brown hessian sack as packets of data, rather than as a shape.

DOI: 10.1057/9781137448859.0005

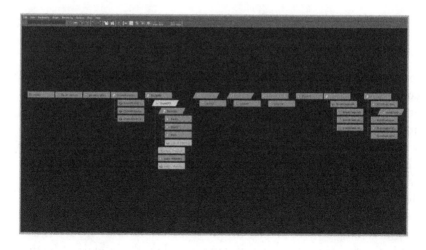

FIGURE 1.4   *Object depicted as packets of data*

Note: The brown sack is seen as connected packets of data rather than as a shape.

Source: Image ©Paul Hilton.

The possibility space of Maya not only exists via the accessible manipulations of the viewport, but also extends to the datasets visible via dialog boxes. As user's 'play' with the software, their possibility spaces for creativity exist through their engagement with both shapes and packets of data.

The claim that users engage with both shapes and packets of data takes us right to the central dilemma of working with software and its operational logic. Creating, when using the software, depends on how effectively the user is able to get to grips with its technicalities as well as manipulating the objects and their movements. Even so, the marketing material for the software suggests that Autodesk is at pains to emphasize its accessibility for artists, echoing the procedural rhetoric of the viewport. The company annually releases upgrades across their suites of software, which includes Maya, 3DS Max, Mudbox, and MotionBuilder. Until 2014, Softimage was included but is now withdrawn. Of the 2015 release, Chris Bradshaw, Autodesk's senior vice president of Media and Entertainment and Marketing, states:

> Our strongest releases in many years, the 2015 versions put more creative power back into the hands of the artists ... Whether our users are creating effects for a movie blockbuster, a game for one of the new consoles or

DOI: 10.1057/9781137448859.0005

developing a mobile app, the 2015 versions allow artists to focus on what they do best – storytelling. (Bradshaw, quoted in Mori, 2014)

This emphasis on putting creative power into the hands of the artist repeats the marketing surrounding the very first release of Maya back in 1998, then known as Alias|Wavefront Maya. In the demo CD publicizing the new software, the video introduced the user interface as 'designed to keep you focused on your creative work', with the signoff flourish remarking that the UI design 'reduces learning time, enhances creativity and throughput'.[1] The main demo video, Alias|Wavefront Maya 1998, too opens with an emphasis on artists and not the software: 'Animators and digital artists have revolutionized the entertainment industry ...'.

Despite such emphasis on artistry, users still have to work with the software. As described above, through the immediacy of the viewport, the software is set more in the background, whereas in the dialog boxes its operations are more explicit. As Maya has traces of the software's deeper structure on the UI, it can be described as both inward and outward facing, a relay of interaction between a surface of toolset menus and the deeper algorithmic structures where data is processed. Seeing an interface as both inward and outward facing resonates with ideas developed by Alexander Galloway in *The Interface Effect*: interfaces are not windows or doorways simply connecting one world with another, but offer locations where different kinds of logic can co-exist and inform each other. For Galloway, despite 'new media foreground[ing] the interface like never before' (Galloway, 2012, p. 30), our understanding remains grounded in the idea of interfaces as points of transition between different formats or logics. Superficially, this would too seem to be the case with Maya's UI since its toolsets provide a connecting layer between users and software algorithms. But by taking into account the inward and outward facing connections, working at the UI can be understood as holding together two sets of logic:

> The interface is this state of 'being on the boundary'. It is that moment where one significant material is understood as distinct from another significant material. In other words, an interface is not a thing, and interface is always an effect. It is always a process or a translation. (ibid., p. 33)

Galloway's framing of the interface as process rather than thing suggests a way forward for thinking about how someone encounters the double logic of Maya's UI. Rather than seeing an encounter as only one thing or the other, the process instead involves a negotiation between the

DOI: 10.1057/9781137448859.0005

surface and depth of the software. Users engage with the persuasive logic of both through whole shapes on the viewport and also data packets in dialog boxes. Martyn Gutteridge (an animator at Digimania) describes Maya's UI in a way that reveals his awareness of the two logics when he compares the virtual projection of a solid object to the nodal organization of the interface:

> So those are all graphically depicted and it's like a network, so you have your different icons and they are connected with inputs showing that this is going to this. It reminds me of, mostly of electronics, logic gates and things. All the different tasks that you want to achieve and, they all have their own utility, they all have their own toolbox that you can go into yourself. And that will affect what you're seeing on the screen.

A further description of this double logic is found in the following long quotation from Barry Sheridan, in which he draws attention to shapes configured as data or numbers:

> It means that things, objects, or animations in the computer's world, they are stored as compartments of information. So for example, just a simple cube consists of data saying that it's a cube, connecting to data that says what size it is, connecting to a set of data that says what material it will have on it, and that's the obvious stuff that goes on to make up the cube. But the cube itself is connected to a set of data that says what lights affect it, a set of data that says what happens to it when it gets rendered, does it cast shadows, does it receive the shadows etc? And so that bit of data which says what lights affect it, that piece of data is also connected to everything else in your scene, so quickly you see that all these links build up. And when you go down an animation pipeline it is even worse, because you've got connections to the rig and the rig is connected to do everything.

In moving between two logics, Maya users are involved in a process of translation. They may be more comfortable directly manipulating shapes in the viewport, but they remain in touch with the dialog boxes. In terms of procedural rhetoric, comments such as those of Martyn Gutteridge and Barry Sheridan reveal how a user is persuaded into a double logic of wholes and parts.

The discussion of the UI has so far focused on exploring the operational logic of the UI visible in the procedural rhetoric of the viewport and dialog boxes. When users engage with the UI, however, they not only encounter the procedural rhetoric of objects as shapes or datasets and the expressive processing of nodes and dependency graphs.

DOI: 10.1057/9781137448859.0005

Another active element is the UI's history or media archaeology. Introducing this history relies on thinking about the UI's permeability to time and place rather than its abstraction as a collection of toolsets. It also meshes with a strong strand of software studies which 'show the stuff of software in some of the many ways that it exists, in which it is experienced and thought through' (Fuller, 2008, p. 1). Some of the stuff of software has already been explored, through how it exists and is experienced in moments of use. Below I explore a software's past, or where it has come from, by looking through the perspective of a media archaeology built from a range of different sources: 'Media archaeology rummages textual, visual, and auditory archives as well as collections of artefacts, emphasizing both the discursive and the material manifestations of culture' (Huhtamo and Parikka, 2011, p. 3). Creating an archaeology for software includes putting more detail into the paths through which a seemingly abstract entity becomes part of a meshwork of meaningful structures. By exploring elements of the development history of the software, the marketing and publicity materials associated with its release and also the wider history of 3-D animation software, I uncover the ways history remains active in the present operations of the software. Of particular interest to me is a continuing attention to negotiating between the demands of artistry and technical expediency, which remains a consistent thread running through the past and present of Maya.

A historical perspective on Maya and computer-generated animation

Martyn Gutteridge of Digimania says of using Maya: 'That is the thing about 3-D. It is the art side and the technical side joining up'. Describing 3-D as a joining up of sides points to the distinctiveness of working with computer-generated animation. Technology has always been part of animation, cameras capturing an image of either a drawing or a stop-motion model. Equally, technology may be involved in creating a drawing or model. The difference of computer-generated animation lies in the numerous automated processes making models and movements happen, and so the joined-up-ness goes deeper. The tricky question that keeps returning, and which runs through the discourses surrounding software, is how things are joined up, what tensions remain under negotiation and

DOI: 10.1057/9781137448859.0005

the extent this process of negotiation feeds into how a user approaches and thinks through the task of creating models and motion.

Before coming onto an exploration of how users in the early 2010s negotiate this tension, I go further into the history of Maya to tell more about the competing influences that defined and continue to define the developments of computer animation software generally and also more specifically Maya and its UI. Thomas Elsaesser, writing about the media archaeology of film history, comments: 'cinema has too many origins, none of which adds up to a chronology, but also makes for doubtful genealogies' (Elsaesser, 2008, p. 233). He goes on to say that a history of film is best understood from the perspective of its multiple influences: 'If we...take seriously the multimedia, multimodal dimensions of our sound, image and text machines, we need to speak of discourse networks rather than apparatus, with its suggestions of fixity, ocular alignment and rigid geometries of space' (ibid., p. 238). Discourse networks are understood as made up of interlocking discourses through which knowledge of something is sustained and reproduced, a process both inclusive and exclusive of ideas. Media archaeology pays attention to gaps that are also part of these interlocking discourses, even as they apparently create a coherent and legitimating rationale for thinking about what a particular thing is or does. In the case of 3-D animation software, for instance, we might ask how the emphasis on photoreal animation has entered into a seemingly natural association with the development of animation software. Elsaesser's thinking about film history is useful when approaching software such as Maya; it too has many origins, and proposing a continuous and linear succession for the software's development would miss many facets of its history. Maya mediates the actions of users and Matthew Fuller and Andrew Goffey say that: '... we consider media and mediation as creating a troubling opacity and thickness in the relations of which they are a part, with an *active* capacity of their own to shape or manipulate the things or people with which they come into contact' (Fuller and Goffey, 2012, p. 5). Thinning out this opacity, making sense of how software is active when in contact with its users, benefits from taking into account the interlocking discourses of Maya's history. Indeed, how the history of software is enacted in the present is an active element in the mediating influences of software and the ways joined-up-ness happens.

Looking at the interlocking discourses of Maya's history allows us to ask the question: where does its active discursive and material qualities

DOI: 10.1057/9781137448859.0005

come from? There are all sorts of ways the discursive and material qualities of the software become joined up. Earlier I noted how marketing publicity for Autodesk 2015 places an emphasis on putting creative power into the hands of the artist, which echoed the marketing for the release of Alias|Wavefront Maya 1.0. The emphasis on creativity in 1998 is itself already a repetition. Ten years earlier in 1988, at a roundtable discussion at the Second Annual Walter Lantz Conference on Animation featuring James Linder, Tina Price, Carl Rosendahl and John Lasseter, the four talked about new computer technologies and animation. Their conversation (republished in 2009) reveals a tension between the expertise needed for both building technology and also telling stories. James Linder noted of computer animation: 'it's a technology that now works, that's not new anymore, and that many people have access to. Now is the time to consider computer animation as a filmmaking tool, and to look at storytelling using computer animation' (Linder in Linder et al., 2009, p. 199). Where initial stages in the developments of computer-generated images and animation had been in the hands of computer scientists with access to hardware and who could program, Linder refers to the second stage (circa 1988), which increasingly brought graphic designers and artists into the mix. At the same roundtable event, John Lasseter, who had shown the short *Luxo Jr* to great acclaim two years earlier at SIGGRAPH 1986, called for everyone to acknowledge the importance of the traditions of animation when using these technologies: 'I want to encourage people working with computer animation, be it from their backgrounds in film, animation, graphic design, or even computers: there's a tremendous history of the development of animation and that definitely should be studied. A lot can be learned from that' (Lasseter in Linder et al., 2009, p. 204).

   Across this mixture of thoughts about computer animation software and storytelling, separated by some 26 years, the sometimes-awkward and tension-filled discursive and material relationship between getting technology to work well and good storytelling is apparent. Building the software's innovative algorithms takes technological expertise; using algorithms also requires some technical insight; doing animation entails a different skill set, and demands of each do not always mesh well. Talking about his experience of looking for a job when computer-generated animation was first introduced more widely within the visual effects and animation industries, James Farrington comments that studios were initially reluctant to employ him because, despite his

DOI: 10.1057/9781137448859.0005

experience as an animator, his computer skills were relatively minimal. Echoing Linder and Lasseter he says: 'It was a harder problem to find people who knew how to animate than could use a computer. We now know that computers are really easy to use, and they're designed so that you know you can pick them up quite quickly.' The awkward relationship between technology and artist, or the potential for awkwardness, is not only found in the discursive relations surrounding computer animation; they are materially traced out in the organizational logic of the UI. Looking into the history of computer animation software illuminates where this logic comes from, how it is visible and the ways in which users negotiate with it. Across software history and the present, the awkwardness never is resolved, remaining a constant in the activity of doing computer animation. There *is* a tension between the desire to tell stories and the necessity of working with technology. The more interesting question is how users negotiate their way through the tension, and what that negotiation in turn reveals.

For Jussi Parikka, media archaeologists accept the complex entanglements of the past and present so that 'archaeology means digging into the background reasons why a certain object, statement or discourse or, for instance, in our case, media apparatus or use habit is able to be born and be picked up and sustain itself in a cultural situation' (Parikka 2012, p. 6). Digging into the background reasons behind why Maya was 'born' as it was, got picked up and then sustained itself within a highly competitive marketplace, includes not just looking at what the software initially offered and continues to offer, but also the context in which it was developed. One key alignment between the discursive and material qualities of Maya is through its status as a commercial product. Following initial success in the computer animation design field, the animation software was developed for markets within the creative industries, including visual effects, animation and games.[2] Aimed at the creative industries, the software's promotion makes clear its design for creating images matching the expectations of those industries, placing an emphasis on photorealism and particular styles of movement, as well as simulations of real-world textures and movements to match with live-action. Writing about the state of 3-D animation software in 1998, the year of Maya's release, David Sturman says:

> Most modern commercial keyframe systems are based on the simple BBOP interactive keyframe approach to animation with added features that ease the animation process. At their core, they all have features of BBOP (some

DOI: 10.1057/9781137448859.0005

copied, some developed independently), including hierarchical skeleton structures, real-time interactive update of transformation values, interpolation of keyframes in channels so that different joints can have different keys in different frames, choice of interpolation functions such as linear, cubic, ease-in and ease-out, immediate playback and an interpolation editor. (Sturman, 1998)

The animation toolsets of Alias|Wavefront Maya included all of these elements, allowing it to map onto the expectations of its potential marketplace. Additionally, the official timeline for Maya's history (made available by Autodesk) draws attention to the multi-functionality of the software. The timeline starts with the merger of three companies in 1995 into Alias|Wavefront: Alias, Wavefront and Silicon Graphics Inc. Subsequently, a blended code was derived from Wavefront Advanced Visualizer, Thomson Digital Image (TDI) Explore, Kinemation, Dynamation and Alias PowerAnimator. From these, Maya was born or developed, and its architecture evolved further to allow customization of the UI and workflow, introducing MEL scripting for flexibility and more user control. The merger of companies and code was significant because it brought together a range of existing functionality from previously successful companies, bringing to that mix key next generation elements (digital hair, skin and clothing) under separate development.[3] The range of functionality was a factor in enabling Maya to be picked up; its attractiveness lay in its match to a pipeline pattern evolving within the visual effects industry and animation studios. For instance, a widely used pipeline in the mid to late 1990s combined PowerAnimator for modelling, Softimage for animating and Photoreal RenderMan for rendering.

Enmeshed in this discursive and material history of corporate mergers and multi-functional software was the construction of the user interface, also sometimes referred to as the graphical user interface or GUI. The organization of the UI is important to this archaeology; for many users it is the place in which the relationship between technology and artistry is negotiated. Other discourses come to inhabit the software at the UI too, in particular the idea that Maya is a flexible software. When introducing the newly released software, the user manual for Maya 1.0 makes the point again that the software allows artists to focus on the 'act of creation,' and notes the interface's key role in enabling this focus:

You will soon learn how Maya's architecture can be explained using a single line – nodes with attributes that are connected. As you work through this book, the meaning of that statement becomes clearer and you will learn to

DOI: 10.1057/9781137448859.0005

appreciate how Maya's interface lets you focus on the act of creation, while giving you access to the power inherent in the underlying architecture. (Alias|Wavefront, 1998, p. xix)

The simple statement that Maya's interface lets you focus on the act of creation belies the interconnecting histories leading into the configuration of both the software and the UI. Given its status as a product whose target was aimed for a wide uptake, the configuration of the UI and flexibility of the software was a concern amongst potential commercial users, an issue taken up by David Sturman:

> The main problem for commercial software producers has been how to integrate a diversity of animation methods into a single animation structure, or conversely, how to create an animation system that will accommodate a grab bag of animation methods. 3D Studio approached the problem by providing an easy pathway for plug-ins. Hundreds of third-party developers have added diverse capabilities to the basic keyframe system including dynamics, facial animation, gait control and motion capture. Softimage and Alias|Wavefront have recently completely redesigned their systems to facilitate plug-ins and custom application development, as well as incorporating new animation methods in each release. (Sturman, 1998)

The discourse around flexible accessibility operates on surface and deeper levels of the UI and pulls in too issues around artistic and technically oriented users. Looking first at the surface level of the UI, early reviews commented on the usefulness of widgets and hotboxes to quickly access tools (Sims, 1998). Such flexible accessibility remains important to sustaining the pick-up of Maya more than ten years later. Introductory tutorials in the 2010s continue to promote the UI's flexibility, emphasizing how different users still find their own routes of access into the software. Escape Studio's *Introduction to Maya* training module draws attention to both the wide range of available toolsets and methods for customizing the UI to individual preferences.[4] Books such as *How to Cheat in Maya 2010* (Luhta, 2010) and *Introducing Maya 2013* (Derakhshani, 2012) devote time to explaining the organization of the UI, with catchphrases such as 'You Put the U in UI' (ibid., p. 50). Flexibility also goes much deeper than the organization of the toolsets on the UI for individual users, and this is visible through a thread of discourse engaging with Maya's viability in a competing marketplace. Alias|Wavefront worked in conjunction with several companies when developing the UI, ensuring Maya would become an open software.

DOI: 10.1057/9781137448859.0005

Its openness relies on scripting based on the computer language MEL (Maya Embedded Language):

> MEL stands for Maya Embedded Language. MEL is built on top of Maya's base architecture and is used to execute commands that you use to build scenes and create user interface elements. In fact, every time you click on a tool, you are executing one or more MEL commands. (Alias|Wavefront, 1998, p. 226)

MEL is a kind of meta-programming in the sense that it tells the deeper level node architecture of the software what to do, either through the standard UI toolsets with which Maya ships, or through scripts written in-house programmers within a studio or by individual users of the software. Writing in 1995, Barbara Robertson remarked that having access to open software was one of the key concerns of individual companies in the animations and games sectors. Scripting was not an innovation of Maya, but its implementation was perceived as going further than competitors such as 3DS Max and MAXscript (Robertson, 1995a).

Using MEL scripting, and since 2007 Python scripting too, makes the UI customizable and allows simple repetitive tasks to be automated or streamlined to enable a focus on particular tasks. As a review of Maya 1.0 states: 'The GUI can be extensively configured to any users production needs or preferences. High-end studios can employ technical directors (TDs) to make a GUI just for key frame animation and hide a lot of the Maya's plumbing' (Sims, 1998). The degree of Maya's openness is often attributed to Alias|Wavefront's collaboration with Disney during the development of the software: 'Disney requested that the User interface of the application be customizable so that a personalized workflow could be created. This was a particular influence in the open architecture of Maya, and partly responsible for its becoming so popular in the industry' (Ryan, 2011, p. 315). Its uptake was likely furthered by the range of companies who beta tested the software, many of whom were significant players in mid to late 1990s period of CGI. Blue Sky/VIFX, Cinesite, Dream Pictures Studio, Dream Quest Images, GLC Productions, Kleiser-Walczak, Rhonda Graphics, Square, Santa Barbara Studios and Imagination Plantation were among many of the beta testing customers to support Maya. While production companies welcomed the flexibility offered by scripting since aspects of the software could be tailored to the studio and its particular productions, commentators noting the

DOI: 10.1057/9781137448859.0005

enhanced scripting of Maya 1.0 raise again the problem of technology for artists without the relevant training:

> Scripting expressive characters, for example, is extremely difficult, not to mention unnatural for an artist. Interactive keyframe systems are just the opposite. They allow artists to interact directly with the objects and figures within a familiar conceptual framework. But they become inefficient or tedious to use for mechanical or complex algorithmic motion. Because they are more easily used by artists, the interactive keyframe approach has won in the commercial software market. Curiously enough, as animators are becoming more sophisticated in their use of computer animation, scripting capabilities are beginning to reappear in keyframe systems. The newest version of Wavefront|Alias' [sic] MAYA animation system has a built-in scripting capability that allows animators to tie actions to events, define movement as functions of other movements, create macros and more. (Sturman, 1998)

The description of Maya's emergence from a range of source codes, and its place as a commercial product determining particular toolsets, which also sparked a design and marketing emphasis on flexibility and accessibility, places the software at the centre of several interlocking discourses. There are numerous tensions running through these different discourses: the increasing standardization of software packages in the 1990s running alongside a desire for flexibility, the emphasis on a style of animation suited to commercial animations, as well as the continuing discourse of creative versus technical skills. Seen in this light, Maya becomes not simply a means for creating computer-generated images, but also a discursive and material site with which discourses become enmeshed. As such, software ceases to be an abstraction, and is instead an active element with the capacity to persuade and engage users in a particular discourse. In this way, the UI operates through a form of governance.

The idea of governance and software has been raised more widely within software studies. In *Programmed Visions*, Wendy Chun 'links computers to governmentality neither at the level of content nor in terms of the many governmental projects that they have enabled, but rather at the level of their architecture and their instrumentality. Computers embody a certain logic of governing or steering through the increasingly complex world around us' (Chun, 2011, p. 9). The UI is only one element to software (if it requires a UI), but there are organizational parallels between Chun's comments on computers and the UI. Unlike software, though, which is 'an invisible program that governs, that makes possible certain actions' (Chun, 2011, p. 57), governance is perceptible on Maya's

DOI: 10.1057/9781137448859.0005

UI. It makes visible some elements of the program that governs as it makes possible certain actions as opposed to others, persuading through the processes that define its operational logic. Finally, Maya's UI does not simply sit alongside a set of interlocking discourses, but is permeable to those discourses too. Moreover, these are activated via the UI as the procedural rhetoric of the viewport and dialog boxes persuade through their processes. Governance becomes discernible in frames and patterns observable on the UI.

Go back before going forward

The frames and patterns found on Maya's UI have discursive and material antecedents in the wider history of software development, as well as in the moment and immediate past of the software's release. Since software is often opaque, Matthew Fuller and Andrew Goffey agree with Wendy Chun in remarking that: 'forms of governance gain ground because no one sees them coming' (Fuller and Goffey, 2012, p. 14). Looking back into the history of computer animation software development allows us to see how some elements of Maya gained ground. In what follows, the entanglements of the past and present are uncovered in contemporaneous commentaries about computer animation and also research papers describing innovations in programming and software development. For complex multifunctional software such as Maya, the range of this material is vast, having the potential to encompass modelling, animating, shading and texturing, lighting, rigging and rendering. My discussion focuses on the timing and posing of figures, as it is a multifaceted area from which to continue thinking about the interlocking discourses of Maya's archaeology.

Moving objects and figures are fundamental to animation. One of the key differences of working with software is the process of making an object or figure move. In comparison to cel-animation or stop-motion, a different way of thinking and planning is necessary when using a computer to animate. Though writing about line drawing rather than animation, artist Collette Bangert, who worked with her programmer husband Jeff Bangert, comments of learning more about her processes for drawing as she began working with a computer in the 1970s:

> Now I am beginning to see what a line is about. To see that I can choose to draw little lines, a one big sweep of the arm line, a coiled or an uncoiled line, crossing lines, spiralled lines, decorative lines, random lines, and it's all

DOI: 10.1057/9781137448859.0005

the same line. Where and how these lines are placed and coloured make the drawing what it is, that composition is perhaps the truly difficult element in the making of a drawing. Now I have really to think about what I am doing while drawing in order for Jeff to write a program to deal with what I can do as second nature. This thinking has made the making of the hand work much clearer. We consider each drawing element as an independent element. This is artificial. Yet, this artificiality is precisely one aspect of the use of a mathematical attitude—the separation and isolation of individual elements of a problem. (Bangert and Bangert, 1975)

Bangert's alertness to the ways a computer program mathematically separates and isolates individual elements of a problem is a starting point for thinking about the difference animation software makes to creating an object or moving a figure. It draws attention to a software's participation in actions when manipulating data and also abstracting processes through breaking down a problem of modelling or movement or shading or lighting into its component parts. Of interest too are the ways software puts those component parts together again, introducing new possibility spaces by generating distinct limits and opportunities with which artists can work. In the case of timing, 3-D animation software is based predominantly on a keyframing method, and various algorithms for creating the in-betweened poses mediate the creative work of animators. Algorithmic operations are one of the locations where users of software fold in with the system as they negotiate with the opportunities of their possibility space.

Before I talk about timing and pose through keyframing, some more insight into the early development of computer animation is useful. In these early developments, the interconnected discursive and material traces that cohere in Maya's UI had begun to already emerge. Tom Sito charts the progression of computer animation as a series of intersecting activities. Innovations in the hardware of screens and processors, as well as software, arose out of a network of influences that included experimental artists, communication industries, military funded research into visualization systems and also a community of hackers excited by finding ways of solving problems through writing algorithms (Sito, 2013, p. 3). Though Sito begins earlier, in the 1950s with Mary Ellen Bute's analogue electronic experiments with oscilloscopes, a familiar starting point in computer animation histories is John Whitney Senior, the first artist-in-residence at IBM in 1966. Maureen Furniss describes how Jack Citroen and Whitney shared an interest in abstraction, mathematics and technology, which lead to the latter's residency at IBM (Furniss, 2013). Whitney was especially fascinated by the opportunity computers presented for applying musical concepts about harmony to visual arts of motion, creating *Permutations* (USA, 1968) and *Osaka One, Two, Three* (USA, 1970), and *Matrix I* and *II* (USA, both 1971) at IBM (Alves, 2005, p. 45). Writing about what he termed 'computational periodics', Whitney said for an artist: 'the computer is his [sic] instrument that can integrate and manipulate image and sound in a way that is as valid for visual, as it is for aural, perception' (Whitney, 1975). Around the same time, Stan VanDerBeek took up an artist-in-residence at the Bell Labs creating *Poem Field* with programmer Ken Knowlton (1966–1969). Ken Knowlton also worked with Lillian Schwartz at Bell, who also went onto become a key figure in computer-generated art. Reflecting in 1986 on computer art between 1965 and 1975, Frank Dietrich commentated on the work of Bell programmers including Knowlton:

> Because these artists were not interested in descriptive or elaborated paint-ing, they could allow themselves to relate to the simple imagery generated by computers. Their interest was fuelled by other capabilities of the computer, for instance its ability to allow the artist to be an omnipotent creator of a new universe with its own physical laws. (Dietrich, 1986, p. 161)

These early computer artists and animators, a group who over the years went on to include Larry Cuba, Rebecca Allen, Vibeke Sorensen, and Char (Charlotte) Davies, worked to a logic which might be described as programming art as opposed to programming software. The notion of programming art is traceable too in Charles Csuri's animation *Hummingbird* (USA, 1967). Another artist who explored the possibility of using computers, Csuri, having learnt programming while at Ohio State University, wrote of the thinking behind *Hummingbird*:

> The motion picture *Hummingbird*... utilizes several functions simultaneously, each varying independently as in the 'three-bird scramble' sequence. Each bird was on a different mathematically defined path, while their size was changed sinusoidally – each at its own frequency. Also each image moved from an abstract picture to a realistic bird at an independent and non-uniform rate. (Csuri and Shaffer, 1968, p. 1295)

Running in parallel with abstract animation, where artists exploited the potential for computer programming written on the basis of separating and isolating individual elements generating novel images, was the aim to accurately visualize something in the world. Where Csuri's comments

DOI: 10.1057/9781137448859.0005

about *Hummingbird* are suggestive of a formal dis-integration of the multiple elements of the algorithmic bird, visualization increasingly sought to integrate the digitally created elements into imagery that more directly mapped onto physical reality.

Accordingly, the aesthetic possibilities offered by 3-D computer animation have tended to be dominated by work approaching the simulation of human movements, photoreal textures and animations that developed into the character-based traditions now exemplified by Pixar's output. Interestingly, figures associated with programming art were also part of programming software's history. Larry Cuba, while at the Electronic Visualization Lab, University of Illinois, created the death star computer graphics for *Star Wars IV: A New Hope* (USA, 1977); Rebecca Allen, at the New York Institute of Technology (NYIT) in the early 1980s, used the newly developed 3-D keyframing software BBOP, and Char Davies, who joined Softimage in 1985, was key to the company's success. In these early years, before 3-D animation had settled into the lineage associated (though not beginning) with *Toy Story* (USA, 1995), many different groups were involved in developing computer graphics and animation, which allowed for a diverse range of interests to be part of the wider community of computer animation. The movement towards visualization as opposed to abstraction and experimentation shifted between corporate communication companies, design research and development, and military funded research groups whose goal was to have increasingly sophisticated imaging. In the USA, military funding, often through the Advanced Research Projects Agency (ARPA), enabled computer scientists and engineers working in universities (Stanford, MIT, Utah, Ohio State and Illinois), as well as IBM and Bell, to develop many features today's computer-generated animation rely on (Sito, 2013, pp. 46–52). As Ed Catmull maintains: 'ARPA would have a profound effect on America, leading directly to the computer revolution and the internet, among countless other innovations' (Catmull, 2014, p. 5). The innovations included powerful computers, pixel based screen technology, interactive light pens and software developments including wireframe modelling and shading protocols, which remain the basis of computer-generated animation.

In the mid-1960s, the possibilities for using computers to create art and also animations for educational purposes were still up for discussion (Knowlton, 1965; Noll, 1967). Ronald Baecker, in a report on based on his ARPA supported research, describes computers as a

DOI: 10.1057/9781137448859.0005

'powerful aid in the creation of beautiful visual phenomena' (Baecker, 1969, p. 273). The animated process Baecker introduces in his report links back to Ivan Sutherland's Sketchpad developed in 1963, the first software to generate drawing on a screen from a user working with a light pen and tablet. Sutherland noted: 'Sketchpad need not only be applied exclusively to engineering drawings ... The ability to make moving drawings suggests that Sketchpad might be used for making animated cartoons' (Sutherland, 1963, p. 343). Baecker goes further and writes that computers offer an interactive interface for artists to create a stream of images and his paper: 'explains how the computer can be a medium which transforms the very nature of the process of defining picture *change*, of defining movement and rhythm. Dynamic behaviour is abstracted by *descriptions of extended picture change*' (Baecker, 1969, p. 274) (emphases in original). The excitement of interactive animation lay in quickly seeing the image on the screen as well as any movements and manipulations.

Baecker's comment, the computer is a medium with potential to transform the process of defining movement and rhythm, resonates with more contemporary ideas about software enabling novel possibility spaces for creating movement in computer-generated images. His observation that dynamic behaviour can be abstracted through extended picture change refers to movement broadly defined. Subsequently, the process of defining movement became strongly aligned with figure-based movements, a result of intersecting influences that only began to become coherent from the 1980s. While some larger studios developed their own versions of software, the decade saw a growing reliance on off-the-shelf packages, with Softimage and Alias early successful examples of 3-D animation software. In terms of timing and moving figures, automated in-betweening of keyframes and the development of algorithms to abstract movement and facilitate automation was a key development for computer-generated animation. Associated with the various material innovations of writing these algorithms are numerous discursive threads. They combine an interplay of technology and user, efficient workplace practices and issues relating to simulated and expressive movement.

Scrutinizing keyframing in more detail, I now draw out some of these threads. First introduced for creating 2-D images, automated in-betweening pre-dates the 1980s. Nestor Burtnyk and Marceli Wein developed one of the first keyframe-based computer animation programmes in 1971 when working at the National Research Board of

DOI: 10.1057/9781137448859.0005

Canada. Like Baecker, the two worked with the pictorial image and claimed animators had 'communication difficulties' when dealing with programming:

> An animator's ideas involve mainly pictures and their motion. Thus it is appropriate that the communication of ideas between the animator and computer should be largely through pictures. Our interactive computer-controlled graphics system allows the animator to develop pictorial sequences directly on a cathode-ray tube display, without forcing the animator to become a computer programmer. (Burtnyk and Wein, 1971, p. 149)

Burtnyk's decision to focus on keyframing as a process open to automation, and so solving these communication difficulties, is often attributed to his visit to a conference in California:

> In 1969, Mr. Burtnyk attended a conference in California where Disney studio animators discussed their craft. 'They said there were principal animators and so-called in-betweeners who handled the fill-in, secondary animation', recalled Mr. Burtnyk, who lives in the Ottawa suburb of Kanata. 'Well, I never had an artistic inclination, but I came back to Ottawa thinking the computer could serve as an in-betweener and help animators fill in the holes to their work'. (Guly, 1997)

Working with keyframes or key poses has become predominant in computer animation. But its lineage goes much further back than the beginnings of software development: 'The 2-D approach is often a direct simulation of the cel process whereby an animator draws the extremes and a computer calculates the intermediate shapes and positions' (Pocock and Rosebush 2002, p. 261). In simulating a cel-process, keyframing sets up a direct lineage with traditional animation. The origins of keyframing are often connected with the Disney Studio, an inaccuracy that Chris Pallant suggests is a likely consequence of the 'principles of animation' published by key figures at the studio (Pallant, 2011, pp. 18–19). Keyframing, by practice if not by name, pre-dates Disney, with the earliest examples often attributed to the McCay split-system used by Winsor McCay on *Gertie the Dinosaur* (1914), and the introduction of cel-animation at the Bray Studio in 1916 (Canemaker, 2009, p. 100). Though initially known as 'extreme positions', the notion of in-betweening and key poses was already a working practice by the later 1910s, becoming more widely established across the 1920s. In traditional cel-animation, pose to pose or keyframe animation remains a predominant practice, though straight ahead animation is also used. Superficially, keyframing

DOI: 10.1057/9781137448859.0005

in computer-generated animation maps onto hand-drawn keyframing. The animator creates a pose, but rather than an in-betweener filling the frames between, software interpolates from one pose to the other. Though sounding simple enough, to make timing happen, the creative and technical skills of an animator fold together with the algorithms of the computer.

Two discourses are often entwined in the development of keyframing. In relation to the ability of computers to reduce the tedium of the production process, Ed Catmull stated in the late 1970s: 'One of the inescapable conclusions one draws after examining the process of conventional animation is that there is a good deal of tedium involved. This observation has led to the idea that the computer can be used to greatly speed up the process and make it cheaper. It is time to examine this idea' (Catmull, 1978, p. 348). Though Catmull was finally more engaged with 3-D animation, his comment echoes an earlier statement by Pierre Moretti at the National Film Board in Canada, with whom Burtnyk and Wein had formed a collaboration to showcase the possibilities of their program:

> 'The animator is attracted to the computer', says animator Pierre Moretti of the National Film Board, Montreal, 'because it is able to handle complex visual structures that would involve a tremendous amount of handwork or which would be impossible to handle by conventional methods ... The hope of lessening the amount of tedious work involved in animation is very interesting to us'. (*Film Animator Today*, 1971)

Pierre Moretti worked with Peter Foldes on *Metadata* (1971), made using Burtnyk and Wein's animation system. Foldes went on to make *Hunger/ Le Faim* (1974), with the latter bringing attention to computer-generated animation, as it won the Cannes Prix du Jury and was nominated for an Academy Award. Watching both films is interesting for many reasons, but especially the '2.5D' look to some of the imagery, especially the face of the character in *Hunger*.

In the end the 2-D in-betweening of this era, based on shape tweening, proved difficult to control, and required a lot of input from animators to fix the flow of movement. As Pocock and Rosebush comment of 2-D in-betweening: 'the goal here, to eliminate the human in-betweener, has proven more difficult than anticipated because often the in-between drawings are not simply a blend of two extreme positions' (Pocock and Rosebush, 2002, p. 261). As these remarks suggest, one of the problems

DOI: 10.1057/9781137448859.0005

faced in the early stages of automated keyframing was in-betweening algorithms operating via an abstraction of movement too rudimentary to simulate those created in hand-drawn cel-animation. These discussions surrounding Burtnyk and Wein's 2-D software bring attention, then, to several discourses assembling around computer animation: to the efficiency of using computers, the awkward relationship between animators and technology as well as the problem of not effectively abstracting movement. These were all to remain active in the emergence of 3-D animation software, which already underway in the 1970s rapidly eclipsed the developments for 2-D.

A *Computer Generated Hand* (1972), made by Ed Catmull for his PhD, was one of the first 3-D animated films. Reflecting back on making this film, Catmull wrote of his experience: 'For the first time, I saw a way to simultaneously create art *and* develop a technical understanding of how to create a new kind of imagery' (Catmull, 2014, p. 13; emphasis in original). By the early 1980s, the potential of 3-D animation technologies were being tried out in several sectors, including advertising, design, visualizations for education and information sharing as well as entertainment. One of the fascinating things about this era of software development is the multiplicity of drivers behind the research and development of movement, which included crash simulation, motion analysis, workplace assessment, dance and movement notation and the culture industries (Badler et al., 1992, pp. 7–8). When creating the increasingly sophisticated software of the 1980s, which stabilized into the off-the-shelf software predating Maya, designers drew on a very interdisciplinary field of knowledge. Commenting on how the emergence of controlling movement in 3-D animation software was informed by numerous disciplines, key figures in the field Nadine Magnenat Thalmann and Daniel Thalmann remarked: 'Historically, we can observe the following evolution: computer animation has started with very simple methods coming from traditional animation and geometry like keyframes. Then, inverse kinematics and dynamics have been imported from robotics leading to complex simulation techniques' (Magnenat Thalmann and Thalmann, 1996, p. 2). A waymark in this evolution was BBOP, the first commercially available 3-D animation software, released in 1983 and created by Garland Stern at New York Institute of Technology. Just as 2-D animation systems had already done, the 3-D animation software operated through keyframing, along with jointed figures, interactive animation, and a virtual camera

DOI: 10.1057/9781137448859.0005

(Stern, 1983). Described as an interactive program, BBOP was aimed at animators rather than programmers.

To give further insight into how 3-D animation both discursively and also materially folds users into its system in particular ways, I look in more detail at movement control and the ways it developed from rival possibilities. When BBOP was released, controlling movement through an articulated figure had already been established as a workable solution to animating bodies. Algorithms were developed to move the joints in relation to one another. Prior to BBOP, in both 2-D and 3-D systems, much research had gone into processes for achieving good movements. This meant both simulating realistic movements for visualizations as well as the more expressive movements of character-based animation. Kinematics became a central element in software from the 1980s onwards, but in the late 1970s, Laban notation was also explored as an option (Badler and Smoliar, 1979; Calvert, 1986). Broadly speaking, kinematics is an abstraction of movement developed for robots, and Laban notation is an abstraction of dance movement. Based on the dance notation developed by Rudolf Laban in 1928, Norman Badler and Stephen Smoliar sought to use the notation system to algorithmically describe the goal and movements of an animated figure. Laban notation relies on configuring the human body as a jointed system, and movements are defined in terms of their goal-oriented trajectories as a set of points in space. It takes into account the constraints of movements in relation to joints and also the extremities of the body. Movement patterns based on using Laban notation are used to specify the animation with a particular emphasis on the end point of the movement. One of the criticisms of algorithms based on Laban notation in the early 1980s was: 'they were good at describing the changes or end results of a movement (*what* should be done), but were coarse or even non-specific when it came to indicating how a movement ought to be performed' (Badler et al.., 1992, p. 9; emphasis in original). Kinematics, whose early developments in computer animation ran alongside Laban influenced programming, relies too on thinking movement through a jointed figure, but derives its motivation from the field of robotics:

> In our view, the control of an animated figure is partly a problem of robotics; we see the animated figure as a simulated robot in a simulated world. There are two major issues involved here. The first is known in robotics as joint-coordination control, or what we call motor control. The motor control problem consists of identifying and implementing, for a given figure, the

DOI: 10.1057/9781137448859.0005

programs needed to coordinate and control the actions of the joints in order
to move the figure. Motion planning, the second issue, involves designing
algorithms for combining or 'blending' motions sequentially or concurrently
and also deriving algorithms for autonomous interaction with the simulated
environment. (Zeltzer, 1982a, p. 53)

In a further article, Zeltzer describes the motor control problem in terms
of goal directed movement and kinematics (Zeltzer, 1982b).[5] Kinematics'
relationship to animation software is via robotics, where kinematics
essentially refers to the motion determined by positions, velocities and
accelerations. Kinematics models motion without needing to account
for its cause. Forward kinematics is used for working out individual
joint rotations in a movement, taken in relation to the hierarchy of a
model and so offers relatively fine control over the movement of a set
of limb joints. A hierarchy is a nested grouping of objects defining the
dependencies of translation, rotation and scale. For a simple leg swing,
a hierarchy runs from hip to knee to ankle, the position of the knee and
ankle follow the trajectory established by the rotation of hip joint (in
terminology still current, the hip is the parent and knee and ankle its
nested children). The potential complexity of this process, both concep-
tually and computationally, becomes clearer if the possible numbers
of degrees of freedom on a human figure is accounted for. Degrees of
freedom refer to the rotational options around a joint. A wrist joint, for
instance, has 2 degrees of freedom, a ball and socket joint such as an
arm socket has 3 degrees of freedom. Even if the human body is simpli-
fied, the articulated figure has 20–50 degrees of freedom, and maybe as
many as 200. Animating one rigid object with six degrees of freedom
for five seconds at 30 frames per second requires 9000 numbers to be
interactively specified.

Despite having flaws too, kinematic algorithms became the preferred
option for animation software; its better fine control over movement
made it a feasible solution. Its introduction coincided with the development of BBOP, often described as the precedent for software such as
Maya. Its implementation in BBOP again draws attention to the tension
between technical and animation skills. In a system such as BBOP,
controlling motion via an articulated jointed figure involves thinking
about the actions of relevant joints in time and space. Avoiding any need
to directly implement algorithms for controlling and co-ordinating the
actions of joints, BBOP made kinematic control and also motion planning or in-betweening more intuitive for animators. Even so, Zeltzer's

DOI: 10.1057/9781137448859.0005

two part model of motion control and planning gives further insight. It describes kinematically defined movements of an articulated figure, and how these fold together with inputs of animators in ways that are different to keyframing based on shape or image. To illustrate, Rebecca Allen, part of the NYIT group in this period, used BBOP to animate the figure of St Catherine that appeared in Twyla Tharp and David Byrne's *The Catherine Wheel* video aired on television in 1982.[6] Allen describes how she used BBOP to keyframe poses based on a videotaped perform-ance of a woman dancing: 'I could study individual [video] frames to determine key positions in particular movements. I then positioned each joint of the computer figure to correspond to a key position in the dance. Approximately two to three key frames were required for each second of motion' (Allen, 1983, p. 38). The computer figure Allen refers to would have been a hierarchical model, whose joints underlay the wireframe for the volumetric dancing figure delineated in the final version by Allen's black and white line design. Fredric Parkes's description of BBOP (and another software TWIXT, released in 1985) illuminates Allen's remarks further:

> The animator can interactively position or pose the animated objects at speci-fied key frames. Software then automatically creates the object positions for all the intermediate frames based on the key poses specified by the animator. The objects or characters being animated typically consist of a hierarchically structured tree of sub-objects. Each node of these structures corresponds to a coordinate system transformation. By manipulating the transformations at each node, the animator can pose the objects or characters as desired. The intermediate frames are created by interpolating the key frame node transfor-mations. In addition to node transformations, the animator can control the camera parameters such as viewing position path and field of view. (Parkes, 1986, p. 90)

These two descriptions from Rebecca Allen and Frederic Parks reveal the thinking behind movement in keyframing and 3-D computer-generated animation, which broadly remain true today. Where both traditional cel-animation and the 2-D keyframing system introduced by Burtnyk and Wein work on the principle of in-betweening through transformation of shapes, 3-D keyframing relies on moving designated parameters within the construction of the entity – often the joints of an articulated figure – which the whole of the entity then follows. Though the integrity of shape still remains paramount, thinking through a movement requires taking into account the hierarchy of the model, the degrees of freedom of a

DOI: 10.1057/9781137448859.0005

joint, and also any character-based qualities that might be embedded in the kinds of movement enabled using kinematics. Writing about software and code, Kitchen and Dodge more recently argue that: 'Regardless of the nature of programming, the code created is the manifestation of a system of thought – an expression of how the world can be captured, represented, processed, and modelled computationally with the outcome subsequently doing work in the world' (Kitchen and Dodge, 2011, p. 26). Such is the case in the developments of 3-D animation software. Kinematics is a system of thought through which movement is captured and modelled. This system of thought then does work in the world by having an influence over the creative work of animators.

Moving towards more recent software, through the 1980s software development continued, with the increasing power of processors adding to the possibilities of computer-based animation. Writing in 1998 for the SIGGRAPH newsletter, David Sturman commented: 'Almost all computer animation today is done using keyframe systems evolved from the early 1980s. Only recently have advances such as full inverse kinematics, dynamics, flocking, automated walk cycles and 3-D morphing made the leap from the academic to commercial sectors' (Sturman, 1998). Though Sturman notes that inverse kinematics and dynamics only became embedded in software in the 1990s, they are worth mentioning in relation to their early development as they further illuminate the possibility space software creates for creating movement. Inverse kinematics (IK) both extends the possibilities of kinematics and simplifies the process, taking away the necessity of separately manipulating each joint in an active part of hierarchical skeleton (Girard and Maciejewski, 1985). Forward kinematics (FK) is effective for basic animations, but is cumbersome to manage, and awkward to use for the nuanced turns and postural shifts of a moving figure. For animators, IK is regarded as more intuitive. Rather than having to place a finger via a top down working through of joint rotations from the arm socket, through elbow and wrist to the finger, the animator simply places the finger and the hierarchy moves into position from bottom up. It should also be added, wholly relying on computer-based interpolations of keyframes when working with either IK or FK would be rare. More usually, varying degrees of intervention are necessary to finesse the movements through manual tweaking or resetting the keyframes.

Another facet of contemporary software introduced in the 1980s was dynamics, which introduced an alternative means for thinking about

DOI: 10.1057/9781137448859.0005

movement. Where kinematics abstracts the motion of an object as movement along an arbitrary path or at an arbitrary velocity (such as turn, roll, rotate), dynamics abstracts motion described by the force that caused it (such as push, shove, pull, drag). Dynamics considers underlying forces and computes motion from initial conditions and physics. Movements based on both kinematics and dynamics include swing, grasp/grip, and twist. For researchers interested in physically correct movements based on collisions and impact, dynamics added a higher degree of realism. Jane Wilhelms suggests that since dynamics are based on physical laws, 'physical simulation has potential for generating complex and realistic motion with relatively little user specification' (Wilhelms, 1991, p. 266). Even though less user specification was necessary: 'the dynamics simulations are driven by forces and toques, which are not a natural way of specifying motion for most animators' (Green, 1991, p. 281). These comments on dynamics again show how algorithms introduce users to a different regime for thinking about movement, which when embedded in software mediate and influence a user's way of thinking.

Looking at some of the commentaries associated with the release of the first computer animation feature *Toy Story* is a useful gauge of the state of play of software in the mid-1990s, during the period of Maya's development. Released to great acclaim in 1995, accounts of the film continue the discourse around the joining up of art and technology already seen in the development of keyframing algorithms:

> 'People will look at Buzz and believe,' he [Lasseter] adds. 'He has rivets and embossing on his butt that says "copyright Disney" that's there because my GI Joe had [embossing]. This collaboration of art and technology is absolutely vital. There is believability because of the technology. (Lasseter, qtd in Robertson, 1995b)

Barbara Robertson describes how *Toy Story* was made using both Alias and also MenV (Modelling enVironment), the latter being Pixar's in-house modelling system. Some 1300 shaders (which both define surfaces and also how they react to light) were developed for the project and allowed the team to create the textures of the toys and locations in which the story unfolds. Similarly, the lighting, which included mood lighting, were effective in bringing visual interest to the surfaces and depth.[7] The description of rigging again reveals a playing off between technical know-how and effectively working with the software. For

DOI: 10.1057/9781137448859.0005

instance, explaining the rigging for the facial animation of Scud, Sid's aggressive dog:

> ... articulation controls were coded into each model, enabling the anima-tors to choreograph action and fit mouth and facial movements to the dialog. Scud, for instance, has 43 controls in his mouth alone to allow him to snarl menacingly and show his fangs. 'Since the controls are going to be used by animators who aren't really computer people, we try to give them models that are easy to deal with,' says modeller Eben Ostby .... (Snider, 1995, p. 5)

Running from the early 2-D computer animation software intro-duced by Burtnyk and Wein, through the developments of various algorithms for controlling movement in 3-D animation, and the off-the-shelf software of the 1980s and 1990s, are a series of interlocking discourses. Connections across this archaeology are far from seamless, often full of gaps, but they hold together enough to reveal a continuing re-iteration of questions around the interplay of technical and creative skills, whether through an attentiveness to intuitive interfaces, the complexity of in-betweening algorithms or commenting on the neces-sity of retaining good storytelling skills. When Maya was released by Alias|Wavefront in 1998, the software was shaped, both directly and indirectly, by the various discursive and material developments in the history of computer animation software. Different algorithms create the possibility space for animators, a space where animators and system fold into each other. The complexities of the algorithms for all the toolsets (kinematics, modelling, shaders, rendering, simulations, dynamics) were minimized through the design of a UI. The emphasis was and continues to be creative engagement rather than assuming a detailed knowledge of programming. Even so, there are many ways in which the mediating tensions of the interplay between technology and creative work remain explicit. Whether in the discourse that surrounds the UI, which frequently attests to the creative fluency of the UI, a UI design actively seeking to draw users under its surface, or the capacity for scripting language to add functionality to the soft-ware, this interplay continues to inform how users engage with the software and how the software is presented to the world. The interface is, then, inhabited by interlocking discourses, a number of which are embedded in the design of the UI. Others have become associated with the software through various kinds of working practices, and

DOI: 10.1057/9781137448859.0005

some run through the marketing, publicity and training materials. As a consequence, working at the UI entangles a user with the past and present of the software.

Working at the interface

So far, my exploration of Maya has looked at how the interface reveals the operational logic of the software through the spatial organizations evident on the viewport: the accessible manipulation of the parameters of time and space and the deeper structures of the software. It has also given thought to the network of interlocking discourses emerging as part of the history of software development, and how they are retained in the configurations of the UI. In this final section, I consider the ways in which these interlocking discourses remain active as a user approaches and thinks through a task. Drawing extensively on interview responses from animators creating moving images for visual effects, animation, games and data visualization, user engagements with these discourses are scrutinized.

Before moving into what animators have to say about working with Maya, there is also the matter of agency. Digital methods can be described as transparent in the sense that they show hide what is going on inside the system. But:

> In creating truly digital methods, we mechanize parts of the heuristic process, and we specify and materialize methodological considerations in technical form. Paradoxically, the practical need to formalize contents and practices into data structures, algorithms, modes of representation, and possibilities for interaction does not necessarily render the methodological procedures more transparent. (Rieder and Röhle, 2012, p. 75)

Though I noted earlier that Maya's operational logic is visible on the UI, to say it necessarily renders methodological procedures transparent is too simple a proposition. The design of the UI deliberately aims to bring together the surface and deep structures of the software, as described by the designers Fitzmaurice and Buxton:

> There is a potential for strong tension between manipulating the surface and deep structure when there is a UI or representational discontinuity between the two structures. One step towards alleviating this is to display both structures and show how the deep structure changes as one manipulates the

DOI: 10.1057/9781137448859.0005

surface structure. Thus, users can build an understanding (perhaps limited) of the deep structure. (Fitzmaurice and Buxton, 1998, p. 67)

The surface and deep structures of the UI are the possibility space within which users create, existing in both the accessible manipulations of the viewport, and also in the packets of data visible via dialog boxes. The central dilemma of working with software and its operational logic is that creating, when using the software, depends on how effectively the user is able to get to grips with its technicalities, as well as manipulating the objects and their movements. Because of this, it is reasonable to suggest that the degrees of transparency of the software can be thought of in terms of gradations. When working with the immediacy of the viewport through direct manipulations of keyframes or shapes, the system remains in the background. By contrast, when working with the dialog boxes and seeing their scene as hierarchical packets of data, users are more aware of the operations of the system. At any given time, users can be operating at either end of this scale or sliding somewhere between. Consequently, transparency might be better considered in relation to more discreet moments of working, rather than an absolute mode of engagement with software.

The idea that gradations exist in the transparency of the system opens up a similar question about agency. Mediation and entanglement are concepts that understand technology to be more than an intermediary, as it changes instead the terms on which actions are carried out. This way of thinking challenges the assumption that all agency lies with the user. Adrian Mackenzie argues agency is distributed, folding in-between people and machines via the mediating influence of software (Mackenzie, 2006). Taking agency to be folded between users and machines offers a contrast with definitions of agency solely referring to that of human users. Nick Wardrip-Fruin, for instance, uses such a definition in relation to operational logic, where he says: 'In short, agency is a term for the audience's ability to form intentions, take actions, and see satisfying results' (Wardrip-Fruin, 2009, p. 344). Agency, in this explanation, has the appearance of being in the hands of the audience or user, whereas Mackenzie's account distributes it between users and algorithms, where the latter is the computational process behind any operational logic. When an algorithm for timing, such as kinematics or dynamics, selects and reinforces one option for action at the expense of others, agency folds in-between the user and software: 'Agency, therefore, is by definition contested in and through algorithms' (Mackenzie, 2006, p. 44). Thinking

DOI: 10.1057/9781137448859.0005

that agency is distributed in such a manner brings into focus the ways in which processes bundle and disperse action between individual and system. The balance between tendencies to bundle or disperse is at the heart of how users experience software.

Exploring this idea further, I look at what software users say about their work with software and the extent to which their insights engage with interlocking discourses permeating the UI. As software is executed, things happen in conjunction with people through an opening up of the possibility spaces. Even when the potential for code to influence is high, working with software remains nuanced and negotiated, its capacity to mediate realized through user practice. Through this negotiation, agency is distributed, bundled and dispersed. Timing is a rich area for thinking through the active qualities of both software and its users. Talking about timing uncovers levels of interlocking discourses, including the joining of art and technology, the varying degrees of mediation involved in keyframing or scripting, the extent to which software bundles and disperses agency and how users engage with both the surface and depth of a software system such as Maya. In addition to its place in the history of software development, timing co-exists in relation to a series of questions about what is meant by meaningful timing. This in turn introduces a range of conventions associated with the contexts in which images are created and exhibited, whether in animations, live-action films, games or visualizations. What is meaningful in one context may not be so in another. In their book *Timing for Animation*, animators Harold Whitaker and John Halas state: 'Timing is the part of animation which gives *meaning* to movement' (Whitaker et al., 2009, p. 2; emphasis in original). Timing is also both straightforward and elusive to describe. While it is easy enough to describe what an animator does, it is another thing to say what makes timing good. This is exactly the problem that lies behind keyframing using a computer. A computer can be 'told' to create a particular series of movements going from A to B, but it is not programmed to nuance the timing on the basis of choreographed movements or characterization. What works in one situation and mood, will not necessarily be of any use in another situation or mood: 'The only real criterion for timing is: if it works effectively on the screen it is good, if it doesn't, it isn't' (ibid.). The parameters for working effectively vary from context to context. For a fight scene in *Kung Fu Panda* (USA, 2008), for instance, animators time for humour, the choreography of the fight, the rhythm of the editing and also the

DOI: 10.1057/9781137448859.0005

different bulk of each of the animal's bodies. A rotund panda will have a distinctive rhythm to that of a sleek tigress. By contrast, the animation for a serious game based on medical triage will need to be accurate to existing medical knowledge. The TruSim division of Blitz Games Studio developed serious games for patient rescue and disaster scenarios, and these relied on showing patient status through changes to skin tone and breathing patterns in avatars that encompassed different genders across a range of ages, races and ethnicities.[8] Data visualizations and molecular movies work to alternative kinds of criteria, depending on whether they are aiming for accuracy or taking a more interpretive visualization. Alex Kekesi of the NASA/Goddard Science Visualization Studio creates animations such as *Seeing Inside a Storm* (USA, 2014).[9] The visualization shows the precipitation patterns of Hurricane Arthur when it made landfall over North Carolina, USA, in early July 2014. Commenting on the process involved in visualization, Alex Kekesi says:

> To us, a data visualization is something that has real data driving it, whereas animations are conceptual in nature. More artistic license can be given with animations versus data visualizations. Data visualizations must always be as accurate as possible. Afterall, our visualizations tell a story about scientific results. Since the scientific result is the core of our story, we are bound to tell that story as accurately and true to the data as possible.

Drew Berry, who makes molecular movies at WEHI TV, explains the extent to which he uses embellishment in his visualizations of molecular events:

> There is artistic embellishment that I use: colour and sound. Colour is meaningless at the molecular scale, as it's smaller than the length of visible light. Choices of colour are very much an emotional engagement tool, so if it's healthy it's going to be pink and red, and if it's diseased it's green or yellow. If it's dead it's cold and blue. Those are the choices, and it just works, particularly for abstract things like molecules, those sorts of subtle cues very much inform how it [the imagery] feels.

Getting timing right, making it meaningful, relies on knowledge not so straightforwardly abstracted by timing algorithms. The contexts for exhibiting moving images, alignments with expressive characterisation or the accuracy of visualization are active influences too.

When animator Navis Binu of Digimania remarks: 'A computer is just a tool and does an interpolation between keyframes provided by animators. It's animators who determine the timing, staging, appeal, personality

DOI: 10.1057/9781137448859.0005

etc., which makes the animation believable and natural,' his comments can be understood to be an intersection of mediating technologies and discourse, which includes the wider context for seeing the final images. Computers (and software) lack a capacity to understand timing; instead, they generate connections between keyframes according to the calculations of an algorithm. Saying the computer does not understand timing suggests that generating the timing of a moving sequence is often a matter of resisting software presence, a practice which bundles agency through both user and software algorithms. Even though software enables by automating in-betweening, equally this creates motion often lacking in nuance and mood, requiring interventions from an animator. Martyn Gutteridge elaborates further on the interplay between user and software when he says:

> And obviously the computer is essentially doing all the in-betweening for you. And as an animator you have to control it, you have to tell it. The computer will just join the dots that you tell it to, that's exactly what the graph editor is basically. That's you telling it how to join up the dots.

Gutteridge's comments introduce an opening for seeing the user and software fold together. How the dots are joined up depends on the algorithm used. The possibility spaces for creating timing rely on the automated processes already described in relation to Maya's archaeology, and many animators note that Maya's algorithms produce a quality of movement that lacks an organic feel. Generally, an animator will want to erase the traces of automation: 'Most of the effort I would say in CGI or animation, it's always taking away the computerness, it is always removing the CG from the CG [Barry Sheridan].' At the same time, in erasing the traces of automation, animators map the movements they craft to the context in which the animation is being created, matching to live-action or cartooning, for instance. The wider context of production is always active in the question of what feels right for any piece of animation and informs an animator's tussle with software to get good timing, perhaps whether they use FK or IK in combination with some dynamics, or not. For instance, when planning and making photoreal cinematics for online games, Aaron Chan of Blizzard Entertainment comments: 'We study the physicality of movements so we are bound to that reality. But we also embellish and push to make it more exciting.'

Animators have diverse approaches to managing movement, some working primarily with keyframes, and tweaking the position of a hand

DOI: 10.1057/9781137448859.0005

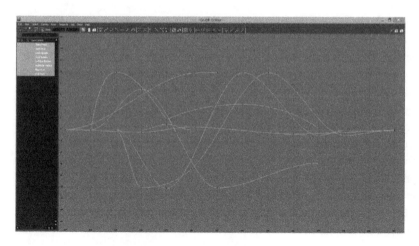

FIGURE 1.5   *Graph editor showing animation curves*

Note: A screenshot of the graph editor showing animation curves.

Source: Image ©Paul Hilton.

or finger or elbow. Others work with keyframes alongside the curves depicting movement in the graph editor (Figure 1.5).

The graph editor is a part of the UI used for manipulating curves of movement and its dialog box takes a user closer to the deeper structures of the software. Sometimes described as the least three-dimensional element of animating, the graph editor displays two-dimensional curves depicting movements in time and space. James Farrington describes his use of the graph editor:

> I personally like working in levels of curves, working to the controls of something. So you know a block will start off with basically just translating it around the screen, and sort of seeing the timing of how long it will spend in each place and then working through the character, and the levels of how it will move, and layering it until you're kind of working at it from a global positioning, right down to fingers and links and tiny detail. I tend to find that what happens with working like that is that it doesn't look very good for a long time, and then it's almost like there's a point where you've got enough rotations and translations sort of keyed in there, so that it will suddenly come to life and then you go through and put lots of detail in.

Unpacking Farrington's comments further, he is talking about working with a figure based on a hierarchical skeleton, where different parts of the body belong to different levels in the hierarchy (Figure 1.6).

DOI: 10.1057/9781137448859.0005

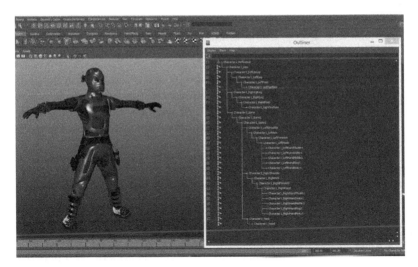

FIGURE 1.6   *Modelled figure shown as a shape and hierarchy*

Note: Screen shot showing a figure on the left hand side and the hierarchy of its grouped elements on the right hand side.

Source: Image ©Paul Hilton.

When talking about controls, Farrington refers to active elements of a hierarchical skeleton rig, or handles that move linked or grouped parts of the skeleton. Controlling movement using a hierarchical skeleton rig means moving only parts of the whole, but these parts are re-connected to the whole through the software processes. The process of controlling movement involves a distribution of agency between user and software. The software disperses user agency in the sense that it establishes the possibility space for movement: a rigged skeleton with movements controlled by automated in-betweening. At the same time, user agency is bundled as they design the kinds of movements a skeleton will have according to the needs of the project. The movements are then created by the animator according to the drama of a story, the promotion of a product or the accuracy of data visualization. In a studio set-up, the rig will be built by a rigger and movements crafted by an animator.[10] Thinking again about Colette Bangert's remarks made in 1975, computer programs mathematically separate and isolate the individual elements of a problem, and this logic is in play with timing. The component parts of a movement are abstracted through the software, broken down into the movable parts defined by the controllers on the skeleton. Putting

DOI: 10.1057/9781137448859.0005

the movements back together into the seamless whole of timing involve a user connecting with the operations of the software and a range of discourses. Animator David O'Reilly observes of using software:

> It is possible for Maya or any 3-D software to feel embedded in the user's mind, so that any conceivable three-dimensional phenomenon can be broken down into a chain of simple actions which eventually reconstruct it digitally. This is perhaps the most interesting side effect of becoming familiar with the software.

Animators who work with the scripting language MEL provide an opportunity to see agency distributed in a different way. Through scripting, users can automate repetitive processes and/or add novel functionality to meet the needs of a specific project. One of the features that made Maya attractive on its release was its level of openness. For users able to script, with access to an in-house team of scriptors or who network with those in the online Maya community willing to share their scripts, scripting offers the opportunity for resetting the boundaries of possibility spaces. Though scripting language had already been introduced for other software: '... Maya was the first of the high end products to build the entire product on a scripting language, meaning the scriptor could "drill down" through the application to the operating system level to build scripts or plug-ins so integrated they become indistinguishable from the core product –– a totally new level of flexibility' (Jeffrey Abouaf, 2000, p. 8). Drew Berry emphasizes flexibility in his comments on Maya: 'It's a package that is just extremely open and flexible to whatever. It's not pre-canned in its toolset. The tools are not built ready to achieve a certain outcome. It's more you can devise a toolset to suit your particular need'. Berry's comment again reveals Maya as part of a wider production process with potential to expand the possibility spaces of a studio. This is true of both commercial and also science visualization studios:

> One of the things our Studio likes the most about Maya is the flexibility it provides via MEL. MEL gives us the ability to adapt Maya for whatever unique challenge we face when it comes to visualizing highly complex NASA scientific data sets. Over the years, we have built a library of MEL scripts that allow us to efficiently ingest NASA-related data ranging from flow fields to ephemerides [Alex Kekesi].

When talking about scripting, individual users draw attention to a different aspect of using software. Scripting increases the influence of code, which in turn introduces a different context for distributing agency. My

DOI: 10.1057/9781137448859.0005

earlier comments about poses and keyframing draw on discourses about expressive characterful animation of movements based on the storytelling conventions of animation. Descriptions of scripting, by contrast, call attention to more procedural examples of animation, where movements are described through lists of transformations rotate, translate, scale specified by data:

> MEL scripts are basically good for repetitive tasks, so with a molecular scene there is often dozens of molecules that are moving in some kind of specified way. With MEL you don't have to do that all by hand, so I don't usually do a lot of hand based keyframing. I usually try to do as much scripting as possible to save time [Janet Iwasa].

Rather than the expressive and characterful movements often generated using keyframing, the skill of scripting involves authoring the rules of movement in code and then allowing those rules to be executed as a series of transformations within a hierarchical model. Just as expressive and characterful movements rely on knowledge of particular styles of animation, scripting molecular movements involves drawing on a background of knowledge about such movements and reconfiguring that knowledge through the possibility spaces of scripting.

Though many commentaries on software outline how effectively it enables users to do what they want to do, engaging with software can also be full of tensions and degrees of frustration, where distributed agency provokes an experience of the limits of the system and/or the skills of the user. For Adrian Mackenzie: '... code, the primary material of software, becomes an involuted nexus connecting people, platforms, reading and writing conventions, power, law and creativity, distributed in time and space. If this nexus holds, it ties people together, but not seamlessly, effortlessly or without tensions' (Mackenzie, 2006, p. 91). Liz Skaggs's discussion of her rig describes the effort of working with software when pushing at or stepping outside of the envelope of conventions for movement:

> I worked on Star Wars [the animated series]. This is cinematic animation, so there is, of course, a lot of running, a lot of fighting, and a lot of fast action. You know, when characters are moving slower in Maya, emoting and speaking and some minor gestures with hands, this is very simple. But when a character is like, 'phew-phew-phew', jumping from wall to ceiling, to God knows where, that is really difficult. My rig was breaking all the time. You want create a pose so that the arm is further than it should be, but you of course can't break the skeleton. So this is like telling Peter Paul

Reubens that he has to stick to anatomy that is real. And you can imagine how frustrating.

Skaggs' comments make clear that negotiation with software is not simply about being able to get it do exactly what one wants. Software offers possibility spaces, but those may have boundaries founded on a relatively inflexible set of parameters. When confronted with the limits of those parameters, the seamlessness of working with software becomes forced and the confines of a software's influence more marked.

Understanding how Maya's UI works has the potential to go beyond being able to work against the system when manipulating graph editor curves or creating scripts to extend a possibility space or reduce the work load around repetitive tasks. It also involves having a greater awareness of the how the software packages data. Knowing more about Maya's deeper structures gives further insight into how software more fully bundles a user's agency. Delving more deeply into what the dialog boxes reveal about the protocols of the software makes evident which elements of the software potentially disperse agency by compelling particular ways of thinking or requiring a specific ordering of workflow. Liz Skaggs comments that using software altered her approach to making animation:

> Because this software translates the in-betweens for you, and you need to look at different tools within the software, such as graph editor, or something like that. You are not creating these images in your mind; you are using some other tool set, a graph or boxes or squares or controllers, different things. So it did have an effect on where to place keys, where to place breakdowns for animation ... So from an animation perspective I guess, yeah, it changes the way you think about animation.

Others talk about the necessity of learning how things are ordered within a chain of commands: 'Modifiers happen in a certain order, and if you start messing around with that order, then things that rely on other things happening earlier or later, may not happen [Barry Sheridan].' An awareness of ordering also informs these comments by Ady Holt: 'And that will become second nature, you know not to put certain things next to others in the hierarchy. But to start with, it's hard to get your head round, it is really not clear what is causing the problem'. Such comments reveal agency being bundled away from the user. Remarks about working to a particular order evince a greater quality of dispersal than seen earlier in relation to timing curves or scripting. Across this range of

DOI: 10.1057/9781137448859.0005

examples and insights from Maya users, as Adrian Mackenzie has said, agency becomes contested in and through algorithms, as well in relation to conventions of movement and studio organization.

Algorithms contest agency by determining what operations do and do not occur. In determining what operations can and cannot occur, these algorithms generate a logical space traced out in the schematic diagrams of the protocols, and the arrangements of packets of data, reconfigured and connected through computation. The logical organization of these packets of data is a feature of program languages. Maya is programmed in C++, which is an object-oriented program (OOP) language. Alan Kay, who first developed object-oriented programming with the language Smalltalk in 1972, describes object-oriented programming as 'a bit like having thousands and thousands of computers all hooked together by a very fast network' (Kay, 1996, p. 517). Each of these 'computers' can be thought of as a subroutine that acts on particular packets of data, with the output of one routine communicated to the next, an organization that speeds up computation and also allows for more complexity. Logical space does not rely on locatedness within a given space and absolute dimensionality, but is instead something that comes into being during the execution of a program as it connects smaller interacting elements. Described in this way, such a space at first seems to be both deferred and diffuse, going back to the idea that software is an abstract entity. Equally, such a description brings out the ways in which the agencies of the user are both dispersed and bundled together. A user's input is traced not only through their manipulations of a time line, or the placement of a keyframe, but as the input moves through packets of data generated by the software's algorithms. The ways these packets of data connect determines how 'sentence structures' and 'logical pathways' influence users.

Conclusion

Approaching software has been the work of this chapter. Thinking about software in some of the many ways it exists, how it is experienced and thought through, involves engaging with software at the place where it makes an appearance in the world, or at least in the world of animators. Thinking only in terms of functionally, the UI of Autodesk Maya connects a user's input to software algorithms. But a UI not only gives access to an array of toolsets, frames and patterns of the surface and deeper structures

DOI: 10.1057/9781137448859.0005

of Maya become humanly meaningful here too. These frames and patterns were thought about in three different ways: operational logics, software history, and user engagement, which together reveal software to be inhabited by sets of interlocking discourses. Running between all three ways of thinking about Autodesk Maya, I noted a concern with negotiating between the creative and technical sides of using software. In detailing this more fully, the toolsets, presented as virtually projected space and discrete packets of data, were teased out to reveal an operational logic. Articulating an engagement with shape versus abstract data, the operational logic creates a location for negotiating tensions that gather when creative work is carried out in the context of an automated system. More understanding of Maya emerged from looking into the context of its production and also the history of computer animation software, with a particular focus on algorithms for movement. Finally, insights gained from interviews were drawn on to explore how and where users and software interact, including thinking through the ways in which discourses developed cross the history of software development remain active for contemporary users. Taken together, these three perspectives show that what software does is a complex entanglement between the software, its users, as well as its discursive and material histories.

Notes

1 The videos are easily found on YouTube by searching for Alias|Wavefront Maya 1.0.
2 When developing PowerAnimator (aka Animator or Alias/1), a precursor to Maya, the company Alias had partnerships with General Motors and Honda. The latter provided financial sponsorship for developing the second generation of the software, Alias/2 (Masson, 1999, p. 214; Design-Engine Education, 2011).
3 For more information about Wavefront, see http://www.creativeplanetnetwork. com/dcp/news/first-wave-origins-wavefront-software/43551
4 Escape Studio is a leading London and LA-based 3-D animation training studio. I was kindly given access to the studio's training videos.
5 In the context of mechanical physics, kinematics is a term first introduced by André-Marie Ampère in 1834 to describe the science of motion, which derives from the French cinematique and Greek kinesis.
6 *The Catherine Wheel*, initially a live ballet performed in 1981, was refashioned for television to include computer animated dancing figures created by Rebecca Allen. The 90-minute video was aired on BBC's Arena arts-based series in 1982.

DOI: 10.1057/9781137448859.0005

7   David Price describes the lighting team as the largest of the technical teams with a list of tasks that included, in addition to mood lighting, time of day and season, guiding the viewer's eye, and underscoring character emotion and personality (Price, 2009, p. 137).

8   Blitz Games Studio ceased trading in September 2013. Their website remained on-line at the time of writing. The TruSim division website (http://www.trusim.com/) included information about the development of their serious games.

9   *Seeing Inside the Storm* is available online at http://svs.gsfc.nasa.gov/goto?11590

10 In the context of a studio, there are organizational structures that allocate agency, such as ones tending to primarily attribute creative input to those working above the line, despite the creative contributions of people working below the line (Stahl, 2005).

DOI: 10.1057/9781137448859.0005

2
Software and the Moving Image: Back to the Screen

Abstract: *Introducing an innovative approach to digital images, Wood explores what software tells us about moving images. Her discussion starts with the logic of realism informing representational spaces in many moving images found in live-action films, games, animations and data visualizations. These are contrasted with sequences found in* Iron Man 3, Oblivion, Rango *and* Journey *that map onto a reality drawing its contours from 'digital space'. Describing digital contours, Wood builds on her discussion of software based on interviews and analysis of Autodesk Maya. Digital contours are explained as movements and proximities in a digitally configured space. Based on non-representational theories used by cultural geographers,* Software and the Moving Image *puts forward more-than-representational space to explain the experience of digital contours in moving images.*

Wood, Aylish. *Software, Animation and the Moving Image: What's in the Box?* Basingstoke: Palgrave Macmillan, 2015. DOI: 10.1057/9781137448859.0006.

Turning back to the screen, in this chapter I shift focus and ask whether knowing about software enables us to approach computer-generated moving images in novel ways. Tracing the difference software makes might well seem straightforward. Pointing to the developments in moving image practices in visual effects cinema, games, adverts, data visualizations, as well as in many feature length and short animations, is easy enough. *Toy Story*, released in 1995, though not the first computer-generated animation, brought the technique much more fully to public awareness. Feted by critics and audiences alike, it set a standard for a particular style of both storytelling and animation.[1] Roger Ebert commented: 'Watching the film, I felt I was in at the dawn of a new era of movie animation, which draws on the best of cartoons and reality, creating a world somewhere in between, where space not only bends but snaps, crackles and pops' (Ebert, 1995). Looking back at *Toy Story* some 20 years later, its digital vintage is apparent. Nevertheless, the textures, lighting and subtleties of movement seen in this ground breaking animation set a precedent whose development is evident in more recent animations, including the *How to Train your Dragon* films (USA, 2010/14), games such as *The Last of Us* (USA, 2013) and *Grand Theft Auto V* (USA, 2013), as well as the visual effects extravaganzas of *Marvel's The Avengers Assemble* (USA, 2012), *Gravity* (UK, 2013) and *Guardians of the Galaxy* (USA, 2014). These developments are visible too in the various digital techniques used in making adverts such as *L'Odyssée de Cartier*, the Cartier advert released in 2012 as a celebration of the 165th year of the company, the first fully computer-generated adverts for Ribena, *Ribena Berries*, as well as molecular animations that were included in the visual elements accompanying Bjork's album *Biophilia* (2011).[2]

   A predominant reaction to these varied images, whether as commentaries or analyses, is to see them as aesthetic creations, representational images including imagined spaces and visualizations of molecules, playable space, or products of a particular set of working circumstances (small or large studios in the visual effects, games or data visualization sectors). Such diverse interpretive frames enable complex and sophisticated perspectives on computer-generated animation, yet the discussion frequently remains on too familiar territory. For instance, emphases on photorealism and representation draw digital entities into a world boundaried by physical reality. As a consequence, the difference software makes is characterized in ways that fit within those boundaries, active in so far as to support these interpretive frames. Wendy Chun, writing

DOI: 10.1057/9781137448859.0006

about software and programming, notes: 'to become transparent, the fact that computers always generate text and images rather than merely represent or reproduce what exists elsewhere must be forgotten' (Chun, 2011, p. 17). Though remaining implicit in Chun's remark, her point that computers *generate* text and images opens up a discussion about what computers, or more specifically software, literally bring to the picture. To tease out what software brings to the picture, I focus on digital media's capacity to reconfigure the relations between things. Digital technologies open images to higher degrees of manipulation, enabling its elements to be isolated and altered, while options for composition in layers have extended well beyond those of optical processes (Belton, 2008; Wood, 2007). With such degrees of manipulation available to image-makers in many domains, including filmmakers, animators, game designers and visualizers, it becomes possible to ask more about the logic of how image elements are combined. Consequently, it is also feasible to examine whether traces of the mediating influence of technology are revealed on the screen. Even though elements constituting an image are frequently arranged according to a logic grounded on physical reality, the claim made here is that images mapping onto a digitally configured space are increasingly visible.

To pursue this idea, I start on familiar territory by looking at the logic of realism informing the representational space of many digit-ally constructed images, whether as simulations of specific events or as representations of physical reality. While the majority of popular films, animations, games, adverts and data visualizations work with this logic, examples discussed include digitally constructed images in films revealing organizations of time and space that draw their contours from 'digital space'. Delving deeper into what a digital space might be, the second section returns to the discrete packets of data encountered in the previous chapter and looks at them in the light of modularity. Modularity is useful insofar as it enables a discussion of the recombina-tion of image elements based on the difference of digital media. Digitally configured spaces, however, can also be approached in terms of how they are experienced or turn up on the screen. An experience of digital configured space relies not only on combinations of image elements, but also on access to those images through movement and degrees of proximity. Digital contours are introduced as an idea for developing my claim further. These are explained as movements possible in a digitally configured space, and to describe them I return in the final section to insights gained from looking more closely at the 3-D animation software Autodesk Maya. Drawing again on interviews with software users, digital contours are given shape. To further develop digital contours, the notion of more-than-representational space is introduced. Based on non-representational theories used by cultural geographers, more-than-representational space is put forward to explain the experience of digital contours in moving images.

Exploring what computers bring to the picture means looking at both representational and more-than-representational spaces. Asking how it is possible to see those reconfigured relations in the images on the screen, also poses a question about our experience of digitally mediated reality. Making sense of the digital is about accounting for its enhancement of depictions and simulations based on conventions of realism, while seeing too how moving images reveal a perspective on the ways mediations of software embedded in the physical world alter our experience of time and space.

Familiar territories

When looking back to the 1960s and 1970s, and the ways in which artists were working with computer-generated images, Frank Dietrich wrote that these artists were interested in the simple imagery created using computers of that era. As he comments: 'Their interest was fuelled by other capabilities of the computer, for instance its ability to allow the artist to be an omnipotent creator of a new universe with its own physical laws' (Dietrich, 1986, p. 161). As computer-generated imagery developed in the decades since, simulation and photorealism came instead to predominate. Remaining a goal in many situations where digital technologies are used in making narrative cinema, this impetus is true too for a number of animations and games. As the visible changes between Pixar's *Toy Story* and *Monsters University* (USA, 2013) show, software makes a difference as the increasingly detailed textures, lighting, and capacity to control movement using motion-capture and animation shows through on the screen. As we saw in the previous chapter, in the making of such images, the algorithmic processes of sophisticated software abstract movement, patterns of light, and configure user inputs as discrete packets of computable data, which are then reconfigured into the images rendered out on the screen. The elements coming together

DOI: 10.1057/9781137448859.0006

and constituting digitally constructed spaces link with the material possibilities of software, defined in terms of the capacities afforded by the functionalities of its algorithms. When thought of in this way, the opportunities for reconnecting packets of data into shapes and space are grounded in the potentialities of the software. There are many different possibilities artists could pursue in imaginatively reconfiguring the relations between things, or creating what Dietrich refers to as realities based on alternative dimensions with their own physical laws.[3] Most times, however, any sense of altered opportunities offered by the distinct materiality of software is covered over, with computer-generated imagery reaching instead towards a touchstone of simulating physical reality. While there are a number of films about computers, few imagine the different materiality of a digitized world. Two examples in narrative cinema and television-based animation remain *Tron* (USA, 1982) and *Homer*[3]. Back in 1995, just as *Toy Story* revelled in bringing computer-generated imagery into line with the traditions of animation in a three-dimensional and recognizable version of reality, *The Simpsons* Halloween special *Treehouse of Horror VI* (USA, 1995) included the segment *Homer*[3]. More so than *Tron*, *Homer*[3] pares back the façade of images on the screen to show off facets of software. In the story, Homer famously crossed over into a digital space through a portal behind a bookcase in the Simpsons' living room and is joined by Bart later in the story. Homer registers the difference of being in 3-D through shock at his 'bulgey' body and awe at being in an environment that he likens to *Tron*. For audiences, the fun is in seeing these very familiar figures in 3-D, inhabiting a space marked out as a grid. Even today, *Homer*[3] remains unusual in incorporating some features of animation software, a translucent grid and primitive shapes with various shaders, into the imagery of its story-world. Two features of the digital environment introduced in *Homer*[3] continue to be relevant for this discussion: the modularity of digital elements (hinted at through the primitive shapes) and the play with a moving virtual camera that pulls back and arcs in space. Modularity moves closer to the idea of discrete packages of data combined in a multiplicity of ways, and a mobile virtual camera enables movements in digital space potentially untethered from physical reality, revealing the mediated experiences of technologically constructed spaces.

Even though the possibilities for reconnecting packets of data into shapes and space are open ended, it remains the case that the distinct materiality of software is frequently concealed, with computer-generated

DOI: 10.1057/9781137448859.0006

imagery organized to simulate physical reality or match to physical reality as it is approximated through the conventions of live-action filmmaking. While audiences may be highly informed about the mediations of digital technologies through various paratextual materials, there is also an expectation that the elements making up an image will be organized in ways that match physical reality, approximating to a logic of realism associated with live-action filmmaking.[4] How this logic holds together both an image and the discourse surrounding those images varies across films, data visualizations and computer games and depends on the context in which those moving images are seen and made sense of. The critical success of a film, for instance, can depend on its audience's ongoing engagement with and knowledge of contemporary narrative cinema. *Guardians of the Galaxy* defied scepticism, negotiating the credibility gap of having an animated raccoon action hero embedded in a live-action film. Partly down to the humorous script, its capacity to persuade relied too on its photoreal visual effects and what Stephen Price has termed perceptual realism (Prince, 1996). The coherency of the Marvel universe was based on many digitally expanded environments, and the complex animated characters Groot and Rocket, whose respective appearances were detailed using textures of wood and fur. In contrast to films such as *Guardians of the Galaxy*, the persuasiveness of data visualizations relies on a different range of collateral material. The Max Planck Institute for Astrophysics has produced several animations depicting the evolution of dark matter.[5] Visualizations such as the *Cluster Movie* (Germany, 2005) reconfigure packets of information derived from computationally modelled projections about dark matter density, gas density and temperature data into simulations. These animations depict the changes as temporal events. Other animations show images from the *Millennium Simulation* (ibid.) and depict the enormous morphological scale of galaxies. The authority of animations like these to stand as plausible simulations of the galaxy reside in the status of excellence given to the Max Planck Institute and the credibility of the computer simulations within the discipline of Astrophysics.[6] They also sit within a lineage of depictions of the universe, which are recognizable to a wider audience and not only astrophysicists, either through documentaries, including *Cosmos: A Space Time Odyssey* (USA, 2014) and *How the Universe Works* (USA, 2010), or variations used in narrative films and animation, such as *Contact* (USA, 1997), *Gravity* (2013), the *Star Trek* franchise (USA, 1979–) or *Wall-E* (USA, 2008). Unless you are viewing such animations

DOI: 10.1057/9781137448859.0006

with specialist knowledge that might contradict or further develop these models, this combination of cues for understanding the simulations of an evolving universe somehow make the imagery seem 'natural'. I use the word natural with caution, as what is natural sits on an intersection of discourses, some of which are outlined just above. Simulations of real world phenomena accumulate status because they translate the data of science into compelling virtual objects. Looking around and beyond these objects is often difficult. Talking about simulations, Sherry Turkle comments: 'Immersed in simulation, we feel exhilarated by possibility... But immersed in simulation, we are also vulnerable. Sometimes it can be hard to remember all that lies beyond it, or even acknowledge that everything is not captured in it' (Turkle, 2009, p. 7). What lies beyond a simulation is as much to do with the means of its creation and software, as it is to do with what has been left out. Commentating on using Maya for data visualization at the NASA/Goddard Science Visualization Centre, Alex Kekesi illuminates the balance between data-driven simulation and creating an effective visualization:

> If we have to create a planet, or an asteroid, or a solar system, we usually have data that will be used to procedurally generate the object(s), rigs, etc. The challenge is how to most effectively ingest the data to create the needed objects and inter-relations between them in our scene. However, once those data-driven objects are created and rendered, we will sometimes get additional insights to what it is we are attempting to visualize. There have been times when the visualization of the data does not intuitively tell the story we're trying to tell, or may even provide further insights into the science.

Where simulations and photoreal images are actively matched to live-action cinema, the various elements that contribute to the coherency of the fictional world keep to a realist logic complementing the traditions of live-action cinema. In such images, echoes of the discrete relations of a software environment are hidden. As the organization of the elements map onto the same laws and physics expected in physical reality, whether in actuality or as shown in moving images, perceptual cueing offers little to suggest other configurations of time and space. The same debt to the organizational logic of perceptual realism can be found knitting together discourses behind computer games simulating the look of live-action films. Produced by Naughty Dog, initially for Playstation 3 and remastered for Playstation 4, the multi-award winning game *The Last of Us* is a good marker of the state of play of photoreal game animation in

DOI: 10.1057/9781137448859.0006

2013. The evolution of photoreal console games is enmeshed in a history of hardware and software development, strategies of game marketing, as well as gamer expectations that the imagery of new games will become more photoreal. Even so, the existing evolution of photoreal gaming has come to stand as a quite natural progression, leaving alternative game-spaces such of that of *Rez* (2001) to seem like an anomaly, at least for platforms such as Playstation.[7]

The iterations of *Grand Theft Auto* (from 3 onwards) and *Crysis* (Germany, 2007–) illustrate how the visual details of console game environments have transformed over the 2000s. The logic of realism goes beyond the environments and character animation in *The Last of Us* extending into the performance of the characters, as well as some of the moral questions posed by its gameplay. Developer Naughty Dog describes the game as narrative driven, with complex flawed characters, and which uses motion-capture of actors to build the animations of the individual characters. It is a dual protagonist game, with the plot and play based around a journey undertaken by Ellie (a 14-year-old girl) and Joel (a man in his late 40s), in a post-apocalyptic scenario where much of the human race have turned into aggressive mutants following infection by a fungus. Ellie has survived an attack by 'the infected', and so she may offer a cure to the infection that has decimated the world's population (though the game is only set in the USA). In each level of the game, the 3-D environments remain grounded in the geometry and physics of reality, with remnants of 2013 depicted some 20 years on as nature erodes the remaining buildings. Despite the difficulty of maintaining real-time rendering for information rich and data heavy scenes, the detail of these environments lies in the lighting, artwork depicting structural decay shaped by encroaching plant-life and on-going battles between competing groups. Enhancing a desolate feel to the visual imagery, the sound design and score relies on delicate sound and quiet, which adds an eerie feel to scenarios. The aim for realism also feeds into character development, which is especially true in the cinematic sequences that focus on playing out the drama of character experiences, rather than moving through the game mechanics of a level. Throughout, the animation is based on motion-captured performances, combined with computer-generated simulations of facial expressions to add emotional depth. In the cinematic that ends the Suburbs chapter, for instance, Ellie and Joel, in tandem with brothers Henry and Sam, hide from a horde of the infected following a skirmish during which, unbeknownst to the

DOI: 10.1057/9781137448859.0006

others, Sam has been bitten. Before going to sleep, Sam, almost 14, and Ellie share their fears about getting from one day to the next and all the losses they have experienced as young teenagers. The next morning, with the fungal infection having taken a grip, Sam attacks Ellie and is shot dead by his older brother, who then kills himself. Played out in the more intense register of dramatic performance rather than action, the sequence aims to deepen the physiological realism of the game.[8]

Running across the digitally constructed elements of *The Last of Us*, from the rigging of facial musculature in Maya, to any enhancements of the motion-captured poses of actors, or the detailed lighting and artwork of environments, as well as the paratextual disclosures providing insights into making and designing the game, is an idea of realism.[9] Writing about the increasingly photoreal imagery possible in real-time gaming, Stuart Andrews, writing in *PCPro*, remarks:

> Photorealism is the holy grail of computer graphics. In movies, whether artists are recreating tigers and raging oceans for *Life of Pi*, or building giant monsters and battling robots for *Pacific Rim*, the goal is to create something that looks real…Even when realism isn't the principal aim – as in *Despicable Me 2* or *Monsters University* – the studios are looking for something to take their film to the next visual level, whether that's through natural textures, realistic fur or sumptuous lighting. (2013)

The emphasis on simulation and creating something that looks real has a long-standing trajectory in computer-generated graphics and animation. From the 1970s onwards, various developments, the rendering of solid objects, Gouraud and Phong shaders, bump mapping and then increasingly sophisticated processes for global illumination, all moved the possibility for simulation forward. While shaders and texture mapping add to the detailing of the imagery, photoreal rendering is central to achieving the look. Noriko Kurachi describes photoreal rendering as a: 'process of image generation by simulating the physical interaction of light on the environment. Specifically, the simulation involves how light is reflected and scattered throughout the environment' (Kurachi, 2011, p. 3). Important as the patterns of light and shadow of photoreal imagery are in contributing to an impression of reality, replicating the perceptual cues of a three-dimensional world is key too. As Stephen Prince remarks: 'Digital tools give filmmakers an unprecedented ability to replicate and emphasize these cues as means for anchoring the scene in a perceptual reality that the viewer will find credible because it follows the same observable laws of physics as the world s/he inhabits' (Prince, 2012, p. 32). Photorealistic digital rendering adds to an image's credibility by being an 'impersonation of indexicality' (North, 2008, p. 22), which gives the impression of 'simulat[ing] photochemical photography's "reality effect"' (Purse, 2013, p. 6). As a consequence, viewers are comfortably able to interpret configurations of time and space from within the conventions of physical reality. Seeing beyond the capacities of mediating technologies to simulate physical reality, to instead transform its contours and our experience of time and space, becomes harder.

As simulations and impersonations of photorealistic imagery accumulate extraordinary detail through shaders, textures and photoreal rendering, interpretations emphasizing mimetic qualities easily continue. So too does an expectation that this will endure in live-action cinema using visual effects, photoreal games, animations and data visualizations. Even so, it is possible to look beyond and find glimpses of other kinds of logic. Much is currently written on the impact of digital technologies on the changing practices of filmmaking, including the styles of filmmaking, ways of viewing and film production (McClean, 2007; Rodowick, 2007; Prince, 2012; Whissel, 2014). And though many writers remain focussed on the impact of technologies in the context of a realist aesthetic drawing on physical reality, there is an evolving body of writing exploring the ways in which moving images reveal traces of their digital origins. Dan North argues that special effects (digital included) 'give the game away and announce their synthetic make-up' (North, 2008, p. 23). Lisa Purse argues that instead of seeing digital technologies as only a tool, we should consider their intersections 'with narrative, thematic and representational structures of a film, and, importantly how [this] registers with the spectator' (Purse, 2013, p. 10). From these scholars we are gaining insight into the ways in which digital technologies contribute to and expand the representational strategies of live-action cinema using visual effects, animation and games. We can go further still, and look towards images beginning to reveal a different kind of logic, one that no longer only maps onto the time and space of physical reality. Some examples exploring this perspective already exist. For instance, Leon Gurevitch refers to the traces of automation in the scale and detailing of computer-generated animation:

> What is apparent in the mass-produced quality of these features, that betrays their synthetic nature, is the fact that nonhuman automation has clearly

DOI: 10.1057/9781137448859.0006

played a large part. The task of rendering and animating so many hundreds of thousands of objects, characters and environments would be so large for analogue animators as to be near impossible. Instead, what the viewer beholds is a composite of animated and simulated image forms only made possible by the synthetic means of computer automation. (Gurevitch, 2012, pp. 134–135)

And William Brown considers the ways digital cinema takes us into 'the realm of the posthuman' (Brown, 2013, p. 48).

In my following discussion of two superhero films, I focus on moments that give glimpses of distinct spatio-temporal organizations. *Iron Man 3* (USA, 2013) and *Thor: The Dark World* (USA, 2014), along with *Guardians of the Galaxy*, are part of the ever-expanding world of superhero films. Just as Marvel's universe has grown through decades of comics since the company was established in 1939, so it continues to be delineated through moving imagery in live-action and animated films as well as games. Superhero films currently rely on an extensive range of effects, and *Iron Man 3* and *Thor: The Dark World* (or *Thor 2*), as sequels and part of the Marvel franchise, meet an expectation of high-end visual effects in their action sequences, a gadget laden environment as well as fantastical characters and worlds. Much of the effects work of both films mimics and simulates the live-action reality of the story-world. *Thor 2* relied on visual effects to create the city of Asgard, Heimdahl's viewing outpost, the alignments of the nine planets, and various marauding creatures inhabiting these worlds, not to mention all the spacecraft. In *Iron Man 3*, visual effects created some forty Iron Man suits for the battle scenes of the finale, including the animation of the MK42 suit, the digital doubles of Iron Man and War Machine (aka Iron Patriot) necessary for action sequences. The destruction of Tony Stark's Malibu mansion, with the building falling into the sea below, was also achieved using computer-generated images.

Even as such imagery predominates, there are also traces of an alternative configuration of time and space. In *Thor 2* the creation of the city of Asgard combines modelling based on the photogrammetry of live-action images of a chain of islands, the Lofoten, in Norway, with the imagined version of the environment separately modelled. In amongst these careful architectural constructions, *Thor 2* has moments that delineate a spatial logic, whose access is explicitly mediated by technologies. One such space is the 'Soul Forge'. Within the narrative, Jane, a scientist and Thor's girlfriend, has been infected by an alien force called the Aether, and having travelled to Asgard with Thor, is taken to a

DOI: 10.1057/9781137448859.0006

medical facility when her actions cause Thor to think she is ill. As Vince Corelli, of Luma Pictures, describes their visual effects-based depiction of the infection, he notes the different logic of the imagery: 'It was like a cosmic MRI that created a three-dimensional representation of Jane's being' (qtd in Fordham, 2014, p. 18). This representation configures Jane's body as particles of molecular energy coalescing into a cloud pattern that roughly defines the shape of her body. Achieved using Houdini and Maya to build a floating volume consisting of swarming particles, the impression is of a body of small elements rather than a continuous entity (Failes, 2013a). The imagery offers not only an imagined depiction of alien scanning technology, but also a re-imagining of spatial relations to define a human body as swarms of mobile points of energy.

Iron Man 3 too has moments that reveal a different spatio-temporal logic configured through mediating technologies. As a sequel in the Marvel Iron Man franchise, the narrative is driven by the battle between Iron Man/Tony Stark and a villain, in this case Aldrich Killian, along with the sub plot of Stark's relationship with Pepper Potts. The place of technology is two-fold. Within the story-world, Jarvis and the Iron Man suits aid in maintaining the narrative trajectory. In terms of filmmaking, the various modelling, compositing, match-moving and animating software, create the integrity of a physical reality inhabited by a superhero figure, and a villain whose metabolism had been altered to explosive ends. In amongst such examples of visual effects, versions of space whose logic is not tied to physical reality but to the functionality of a digital system are also visible. This is especially true of the virtual crime scene. When Stark re-visits the explosion at the Chinese Theatre via virtual projection, the holograph plays out a spatio-temporal logic that owes more to the idea of a database than durational time. The sequence opens up two kinds of configurations, one in which time and space is abstracted into stacks of information and another that is digitally organized. Depicting a timespace abstracted into information, a mobile circular array of nested stacks of data files is projected at around head height. These open up to reveal discrete pieces of information chronologically detailing events, where continuities and differences are configured into a virtually projected database. For Lev Manovich, database logic is one of the defining forms of new media (Manovich, 2001). Instead of a narrative organization of information, the latter is instead configured as database. Interestingly, the imagery of the crime scene reconstruction places both database and narrative logic alongside one another. The technological

DOI: 10.1057/9781137448859.0006

interface displays the database logic, while Tony Stark carries the narrative logic, as he seeks to establish a linearity for all the information he has on hand.

The virtual crime scene reconstruction reveals another technological mediation of space, in addition to the two outlined above. This makes visible a time and space that draws on digital configurations. Superficially, as it is depicted as a projection that diagrams the explosion, including any humans and objects within the space, it gives the impression of mapping onto physical space (Figure 2.1). But, the images shift, both as content relating to the event (the explosion), and as modulations in the relations between the elements that constitute the images. For instance, the giant Heaven Dog shifts from higher to lower degrees of resolution indicating a computational temporality. With time and space delineated in this way, the relations between the elements making up an object are no longer configured into a simulation of physical reality. Instead, on offer is an alternative perspective on information, and the different modes through which it can be experienced and engaged are on show. The virtual crime scene reconstruction is generated from a multiplicity of spatio-temporal relations, a mediated experience of time and space where its computational is logic explicit.

Comparing the virtual crime scene reconstruction to the earlier holographic projection when Killian virtually projects his brain during his meeting with Pepper Potts provides an interesting contrast. Rather

FIGURE 2.1   *Still from Iron Man 3 showing virtual crime scene*

Note: Still from *Iron Man 3* showing the crime scene reconstruction virtually projected inside Tony Stark's garage.

Source: Image ©Marvel.

DOI: 10.1057/9781137448859.0006

than bringing the computational form of the imagery to the surface, it remains hidden as it is again mapped onto physical parameters, those of a human brain. A prompt to think in terms of physical reality opens the holographic projection. Killian accidently begins his demonstration with a projection of the universe, providing a brief link to a recognizable version of physical reality before switching to an image of his brain. Like Stark in the virtual crime scene, Killian and Potts stand inside the projection, whose parameters were based on patterns of activity running along the fibre pathways of a brain. As described by Jason Bath, the effects producer involved in creating the image of the hologram:

> 'The brief for the brain hologram was not to have an instantly recognisable representation,' says FUEL producer Jason Bath. 'It needed to be a more ethereal and beautiful interpretation so we looked to stay away from any "traditional" depiction of synapses. Instead we researched modern DTI (diffusion tensor imaging) scans of the human brain, and with the help of data supplied to us from university researchers, created a base CG model of the brain's fibre pathways. We used these curves to animate light moving within the brain. The outer surface of the brain was a separate model with different shaders, and its visibility and lighting intensity were constantly shifting.' (qtd in Failes, 2013b)

Creating this projection of Killian's brain in *Iron Man 3* involved a complex rendering procedure known as deep imaging or compositing, which allows filmmakers to achieve a nuanced interplay of composited computer-generated and live-action elements. Deep compositing, also used in *Prometheus* (UK, 2012), *The Hobbit: The Desolation of Smaug* (NZ, 2013) and *Dawn of the Planet of the Apes* (USA, 2014), enables careful placing of a piece of visual information in space. To get a better sense of what this involves, imagine a view into the depth of a scene, which includes transparent elements or a cloud that can alter the transmission of light. Older composting techniques would allow compositors to bake in a single piece of information about light transition along any particular line into the depth. When using deep compositing, an array of numbers is retained, and so information is held not only at one position on the line, but in several points along the depth of the line:

> But with deep comp, instead of one number there is a range or array of numbers that express the intensity of that cloud from in front of the cloud, through the cloud and behind the cloud. With deep comp, we can composite the live action plate 'in' the rendered 3-D cloud since we can look up in this

DOI: 10.1057/9781137448859.0006

array of numbers exactly how much smoke would be in front or behind at say 18.2m or 16.1m etc. (Seymour, 2014)

This description of deep compositing resonates with the computationally informed spatio-temporal organization noted in relation to the virtual projection of the crime scene. For any given time, the data holds a multiplicity of space points. The final image becomes a question of choice rather than emerging as a consequence of continuity. Unlike the fictional crime scene projection, such computational abstraction of time and space is not usually revealed in the story-world. Indeed, Killian comments that the projection of his brain is: 'Strangely mimetic though, wouldn't you say?'

Thor 2 and *Iron Man 3*, though predominantly working with the reality effect of narrative cinema, include moments where a different logic appears, and relations between elements making up an object are not configured into a simulation of physical reality seen in animations of the universe or *The Last of Us*. Instead, the continuity of time and space is reconfigured, opening them up to a different kind of organization and experience. Central to such a reconfiguration is a changed understanding of the relations between elements that are used to create and build the visible objects on the screen. Knowing more about software provides a way of exploring how this is mediated by the possibilities of a digital environment. Developing this claim further, I particularly focus on modularity and its relationship to the discrete packets of data underpinning the operation of the software and execution of its algorithms.

Under the surface

There are different degrees to which it is possible to get under the surface of the screen, from paring back down as far as program language, to thinking about ranges of toolsets, or understanding the deep and surface structures of software such as Autodesk Maya via its interface, an approach developed in the previous chapter. This discussion continues working with insights gathered from this latter approach, but first briefly considers object-oriented programming before coming back to ways of thinking about moving images.

Autodesk Maya is programmed using C++, an object-oriented program (OOP).[10] In *Unit Operations* Ian Bogost describes object technology

DOI: 10.1057/9781137448859.0006

as a 'set of techniques for constructing software from re-usable parts' (Bogost, 2006, p. 39). To briefly outline, object-orientation consists of four properties: abstraction, encapsulation, polymorphism and inheritance. Taken together these create a form of programming based on the interaction of discrete units of code.[11] Abstraction refers to an object represented in code, which is extracted as a set of shared commonalities. Encapsulation is a process whereby data elements are bundled together and hidden from other parts of the system. It creates discreetly coded structures or objects that can interact through 'messages' running between the objects. In this sense, object-orientation is interested in the relations between things. Rather than computation operating through strings of one-at-a-time instructions, it functions through interactions between bundled data elements in the encapsulated objects where the ordering of events is not rigidly described. Because of this, OOP is sometimes described as a non-linear way of organizing discrete packets of code. Taken more broadly, software systems model things, relations and events in the real world, and OOP is frequently understood as a particular way of directly translating real world objects and interactions into code. Matthew Fuller and Andrew Goffey suggest that this neutral view of OOP can be challenged; they argue: 'computer programming involves a creative working with the properties, capacities and tendencies offered to it by its environment that is obscurely productive of new kinds of entities, about which we know very little' (Fuller and Goffey, 2014, p. 226). As we saw in the previous chapter, Maya operates through a nodal organization, with hierarchies and grouped packets of information co-ordinated to bundle data elements. Bogost differentiates between a process of condensing the complexities of the natural world into mathematical reductions and encapsulating representations of the real world into specific software structures. The encapsulated objects of OOP operate as software structures whereby: 'the logical structures of software design have begun to remap themselves back onto the material world they were invented to represent' (Bogost, 2006, p. 40). Such a remapping is evident in program elements designed to enable movements, such as kinematics and dynamics, where motion is abstracted and reconfigured through the structures of the software. In being reconfigured through and folded into the structures of the software, this remapping is also adding something back that was not there in the first place, something that originates in the execution of the algorithm. Fuller and Goffey's observation about OOP resonates with my discussion of

DOI: 10.1057/9781137448859.0006

digital traces in moving images. Though wanting to avoid a too literal translation of obscure digital entities in OOP programmed software to digital contours, the ideas share a connection. In images where traces of a digital environment are present, a different spatio-temporal logic lies behind the organization of those elements. A shared pattern of organization found in digital environments is that of discrete packets of data (as opposed to shapes), discrete bundles of encapsulated code and discrete modular elements that build into larger units. In the previous chapter, I explored how software dealt with information by thinking in terms of discrete packets of data. My focus was on automated processes in a program and how these joined up packets of data. Taking this idea further, the view that a program participates in the joining up of data, re-modelling the relations between elements making up the images on the screen, is the basis of digital contours. As I argue more fully later, the more-than-representational space of digital contours give rise to the traces of digital spaces on the screen.

Approaching images created using digital technologies as discrete packets of data, or modules of information, is not altogether new. Within cinema studies and what (already) used to be called new media, ideas relating to the time and space of digitally constructed imagery have been influenced by the work of both Lev Manovich and also D.N. Rodowick. They have drawn attention to the discreteness and modularity of digitally constructed imagery and considered the implications this has for encountering the image.[12] As their approaches share a starting position with mine by acknowledging the discreteness of digital imagery, it is useful to go into a little of the detail of their arguments. In the end, Manovich, to an extent, and Rodowick in particular, remain focussed on the discreteness of digital image elements. They emphasise digital discreteness versus continuity in analogue cinema in their conceptualizations of temporality. By contrast, I put forward digital contours as a way of thinking about how we come upon the spaces of digital images differently.

Lev Manovich noted the discreteness of digital elements in *The Language of New Media* when he reasoned that any new media object has the same modular structure throughout. Broadly defined, modularity refers to the ways in which a system's components have the potential to be separated and recombined. Underpinning Manovich's view of modularity is the idea that media elements consist of discrete parts:

> Media elements, be they images, sounds, shapes, or behaviours, are represented as collections of discrete samples (pixels, polygons, voxels, characters,

DOI: 10.1057/9781137448859.0006

scripts). These elements are assembled into larger-scale objects but continue to maintain their separate identities. The objects themselves can be combined into even larger objects, again without losing their independence. (Manovich, 2001, p. 30)

Modularity is pertinent to thinking about the composition of shots and images. From the perspective of modularity, in a digital image any of the many elements in a frame can be identified, pulled out and changed in some way. The virtual crime scene reconstruction of *Iron Man 3* illustrates the process of modular organization. At the end of the reconstruction sequence, Tony Stark, having identified and isolated the specific element that interests him, literally picks it out of its surrounding context using data gloves and scrutinizes it for the information it contains. Echoing the protocol of an archaeological dig, the scrutiny takes the form of brushing back surface information, a virtual dusting away of debris to reveal the identity tags of a man who had been at the Chinese Theatre when the explosion happened. When Stark picks out this modular element, shown as a rectangular cube of data, its excision has no impact on the surrounding information. The base of the virtual projection is interesting to dwell on further. On the one hand, it depicts the famous walk of fame slabs located by the Chinese Theatre, on the other these are grid-like structures echoing those typically found in the viewports of 3-D animation software, Maya included. This grid pattern was evident in *Homer*[3] when Homer and then Bart entered into 3-D digital space. Even as the crime scene reconstruction offers a nicely literal instance of modularity within the context of a film, the sequence also reveals another facet that will become important to this discussion. Digital elements revealing the alternative configurations of digital space only rarely exist in isolation from more conventional ones. More usually, they co-exist with digital elements whose organizations map onto physical reality. In the crime scene reconstruction, in addition to the grid-like structure, the background of Stark's workspace, often seen as a set within the series of *Iron Man* films, was digitally added to allow for the complex match-moving and lighting:

> After the bombing at the Chinese Theater, Tony Stark investigates the site using a holographic re-construction of the scene. 'We had to track the camera and rotomate the actors in some cases – especially when Stark was moving his hands to control Jarvis,' says Bath. 'We built all of Stark's garage in 3-D because we needed to be able to have our holograms intersect and collide with those objects (there is a LOT of CG lighting interaction with the live

DOI: 10.1057/9781137448859.0006

action set). The garage model had to be exact because we knew it would all have to track – and stick – across all of the various angles.' (qtd in Failes, 2013b)

Since the virtual crime scene reconstruction scene aligns two kinds of logic over one another, it has the effect of pulling the image apart, of revealing its modularity. Even though the live-action and perceptually real elements have the appearance of being coherent, they can be modified and manipulated by a range of digital techniques that disaggregate the image.

When thinking in terms of modularity, an emphasis is placed on the relationships between elements within a shot, which enhances a sense of their ability to be separable, their stand-alone-ness. In his philosophy of cinema, *The Virtual Life of Film*, D.N. Rodowick further considers modularity and space, but bringing too a perspective on time. Comparing what he calls a digital event with an analogue shot, Rodowick sees digital images as singular combinations of discrete elements. His elaboration echoes Manovich's explanation of modularity in media elements more broadly:

> A digital event [equivalent to a shot], then, is any discrete alteration of image or sound data at whatever scale internal to the image. Elements may be added, subtracted, or refashioned interactively because the data components retain their separate, modular identities throughout the 'editing' process. (Rodowick, 2007, p. 167)

Both Manovich and Rodowick draw attention to the ways in which discrete packets of data computed into imagery via particular software can be manipulated, modified, separated and recombined in a digital environment through the operations of that software. By virtue of their status as modular, and because they are no longer always secured in the particular spatio-temporal relationality of earlier analogue media, digital objects come to be understood in terms of a changed set of relations between things. Rodowick illustrates his argument through a discussion of compositing and editing, which is helpful to consider further here too because it more fully details his understanding of the temporal relations of digital elements. Because they are made of parts, digital images can be composited and manipulated in layers. As described already, in most narrative cinema, assembling layers and their component elements conform to the cues of perceptual realism. Given smooth continuity and seamless boundaries, the capacity to

DOI: 10.1057/9781137448859.0006

disaggregate elements of the image remains hidden behind a coherent composition, such as those discussed in *The Last of Us* or the effects sequences in *Thor 2* or *Iron Man 3* that map onto physical reality. Despite such a tendency towards perceptual realism in narrative cinema using effects, many games and animations and also data visualizations for Rodowick no longer have the same quality of integrity found in an analogue image: 'space no longer has duration; rather, any definable quality of the defined space is discrete and variable' (ibid., p. 169). Since Rodowick is engaging with a long-held philosophical tradition within cinema studies, which is to explore the connections between duration or *durée* and its expression across the history of film, the difference between analogue and digital images matter. Analogue shots, unlike digital events, consist of spatial wholes. The elements making up their composition remain inseparable, with combination only possible through editing together patterns of those spatial wholes. The integrity of the elements, the relations between everything in the time and space of the shot is always intact, they cannot be 'cut into' in the same way as may occur for a digital event. The emphasis on modularity in digital images, and discrete elements more generally, creates then a perspective on the spatial relations that generates a sense of stand-alone-ness. Such relations focus on interchangeability and/or the fabrication of larger units, with all expressions of digital cinema standing as examples of montage: 'But here montage is no longer an expression of time and duration; it is rather a manipulation of the layers of the modularized image subject to a variety of algorithmic transformations' (ibid., p. 173).

With such an emphasis on manipulable modularity and the missing qualities of continuity essential to conceptualizations of duration, we are left with a sense of lack in a digital event, its not-ness so to speak, which leaves open many questions about what can be more productively said about the digital image. Rodowick aims to get out of the impasse by arguing that the temporality of the digital image is to anticipate a future and can be seen as an expression of a different kind of reality: '... synthetic imagery is neither an inferior representation of physical reality nor a failed replacement for the photographic, but rather a fully coherent expression of a different reality ...' (ibid., p. 176). This interpretation draws on Lev Manovich's position that computer-generated images, or what he refers to as synthetic images, are too real, the consequence of a different or more perfect than human vision:

DOI: 10.1057/9781137448859.0006

They are perfectly realistic representations of a cyborg body yet to come, of a world reduced to geometry, where efficient representation via a geometric model becomes the basis of reality. The synthetic image simply represents the future. In other words, *if a traditional photograph always points to a past event, a synthetic photograph always points to a future event.* (Manovich, 2001, pp. 202–203; emphasis in original)

What is interesting in Manovich's comment is not his rationale that digital imagery is future oriented because of its too perfect vision, predicting the augmented vision of future humans becoming cyborg. Instead, it is the claim that the world is reduced to geometry, a view resonating with Ian Bogost's comment that the logic behind software design structures and frame our experience of the material world (Bogost, 2006, p.40).

As we have already seen, a frequently made point about digital imagery in games or as visual effects in live-action films, or animation, is the way such imagery mimics reality. Perceptual cues map onto the familiar terrain of physical reality, creating depth perspective, gravity, light and shadow. Sometimes this is achieved through highly photorealistic rendering, sometimes not: *Despicable Me* films (USA, 2010, 2013) have textures and lighting recalling a cartoon aesthetic, whereas *Rango* (USA, 2011) has a stronger photoreal look (Figure 2.2). Occasionally, there is a mix of the two. The computer-generated commercials for Ribena Plus, *Ribena Berries* (UK, 2011–), made by MPC, depict photoreal environments (including leaves, grass and meadow flowers, tree bark, fungus) inhabited by realistic squirrels, as well as berries that have a more plastic or cartoon quality. Despite their different textures, all these animations use perceptual cues to give the impression of movements in a three-dimensional world, with shadows and reflection used to enhance depth and a sense of placement within a space that is grounded in a familiar reality. Made using 3-D animation software, including Maya, the worlds would indeed have been broken down to geometry, but put together in a way that covered over its different origins. As such, the logic of the software design and its structuring influence on our experience of the world is opaque. As algorithms improve, as processing power advances, and if the demand still exists, the density of this opacity is likely to deepen, making it harder though not impossible to see the traces of software.

The Lego Movie (USA, 2014) and range of Lego computer games, is an interesting animation to consider in this light, as it adheres to a strong sense of perceptual realism. Even so, it is possible to draw out the traces

DOI: 10.1057/9781137448859.0006

FIGURE 2.2   *Comparison of rendering in* Despicable Me 2 and Rango

Note: Stills from *Despicable Me 2* and *Rango* comparing the different styles of the imagery. Both rely on perceptual realism, but *Despicable Me 2's* (top) cartoon style contrasts with the high degree of photorealistic rendering used in the figures and environment of *Rango* (bottom).

Sources: Despicable Me 2 image ©Universal Pictures; *Rango* image ©Industrial Light and Magic.

of software and questions of its geometry. According to their marketing publicity, the filmmakers wanted to create an environment where viewers could almost believe that the figures were real, and although the animation is almost completely computer-generated, it was made to have a stop motion feel.[13] The environments were frequently based on existing Lego sets, albeit influenced by the works of Finnish photographer Vesa

DOI: 10.1057/9781137448859.0006

Lehtimäki (aka Avanaut) (Failes, 2014). The figures are highly photo-realistic, right down to the details of wear and tear. Their movements were designed to also match the expectations of an audience who have either played with the figures themselves or seen children playing with them. The lighting too is claimed to be equivalent to a Lego construction built in physical reality, though it was most likely tweaked to match with the demands of action-based composition and various animation and cinematic conventions.

Paring back this surface level information about the photorealistic rendering of the figures and environments, Animal Logic's narrative around the making of *The Lego Movie* also touches on questions to do with geometry. When first designing the various figures, the team worked with the virtual Lego Digital Designer (LDD) to mock up models based on the basic brick geometry of physical Lego bricks. Just as the complex dimensions of actuality are abstracted to be workable in software, the same occurred with the bricks. The individual virtual bricks on LDD had a geometry, which when combined into larger shapes produced degrees of redundancy. For instance, the geometry of two brick faces clicked together is not relevant to the modelling or animation of the larger figure or object. Consequently, as the models passed through the production pipeline, the redundant geometry was set aside:

> 'At the very front end it was a brick based approach', says Sarsfield. 'As it moves down the pipeline, that gets baked into more of a standard geometry approach, but we maintain the connection to the brick database – each one of the bricks is recorded in the model dataset.' (qtd in Failes, 2014)

However, if the object needed to be returned to its brick elements, that geometry could be unpacked. During explosions, buildings shattered into constituent bricks, puffs of smoke and fire were broken down into the smallest brick pieces rather than anything more physically realistic. In scenes showing the ocean, the roll of the waves reveals a complex simulation of blue and white bricks, with the texture of the water's surface breaking up into bricks of varying sizes. Paying attention to how the geometry of both the virtual Lego bricks and that of the animation soft-ware used, reveals competing tendencies in using software, between the bigger picture and the smaller units behind that picture. This echoes the narrative of Animal Logic, whose emphasis on the photorealism of the figures and environments tends to draw attention away from the modular basis of the bricks. Even so, modularity keeps breaking through

DOI: 10.1057/9781137448859.0006

the surface, revealing how brick-logic structures and frame the Lego World characters' experience of their material world. Although it is rare for such modularity to break onto the surface of most films, animation and games, *The Lego Movie* is a reminder of its presence behind the scenes in all computer-generated elements.

So far the focus on modularity has paid attention to the relations between the elements that make up the image as a series of representational spaces and entities. A further perspective on digital space comes into play by thinking too about how movement configures those relations for a viewer. D.N. Rodowick discusses movement in relation to modularity, noting:

> [W]e continue to be surprised and disturbed by perceptually convincing viewpoints unanchored by gravity and by spaces that appear to morph, disassemble, and recreate themselves according to an astonishing variety of parameters ... And so, from the perspective of a filmic culture, one of the most unnerving and often thrilling aspects of digital space is the sense of controlled, continuous, and open-ended movement. (Rodowick, 2007, p. 170–171)

The surprising, disturbing thrill of movement in a digital space opens it up to an emphasis beyond modularity and the geometry of shapes. A question can also be posed about the geometry of movement, by which I mean the geometry of movements in the space around digital entities and also the movements of the objects themselves. Talking about movement through digital space benefits from further qualification. Strictly speaking, there is no movement in a virtual space, only the impression of movement conjured by the synthesis of motion perspectives. Motion perspectives embedded in perceptual realism and simulations can be anticipated to map onto expectations set by human perceptions of physical reality. Even though technologically mediated, the contours of movement in cinema, for instance, remain connected to forces associated with physical conditions, especially those of gravity. This remains the case with cameras whose trajectories and movements extend those of hand held or pedestal cameras. Cameras operated using cranes, under water with divers or submersibles, or at height in aircraft or even in space flight, stay allied to the logic of those physical conditions. A camera in a submersible, for instance, can only go where the vehicle goes. With Steadicam, a fluidity of motion and micro-adjustments in position becomes feasible, but is still associated with the physical location and mobility of the camera operator.[14]

DOI: 10.1057/9781137448859.0006

A comparison of some different uses of mobile camera shots in live-action illustrates motion perspectives both connected to and disconnected from the parameters of physical reality. The music video made by OK Go for *The Writing's On the Wall* (UK, 2014) features Steadicam work allied with numerous optical tricks to give the impression of a continuous shot just over four minutes (4 minutes 17 seconds) in length. Shot breakdowns reveal the extent to which the video, made in 65 takes, relied on rigs and makeshift devices to allow the fluid movements between levels of a built structure in a warehouse or around obstacles in the space of the shot.[15] The physical integrity of the space of OK Go's music video, despite the many takes, cuts and tricks, creates a motion perspective that contrasts with the adverts for the Definitive collection of clothes by very.co.uk. Made by St. Luke's Agency, the adverts feature static figures around which the camera moves. It is not so much the camera movements that are unusual, but the conjunction of stillness and motion. The poses of the figures exaggerate this strange quality: the ponytail of a women defies gravity by remaining flicked up in the air, milk pours into an enormous persistent splash, and a heavy ten-pin bowling ball hangs in the air. This is the antithesis of the Steadicam shot used at the end of *Hugo* (USA, 2011). Drawing all the threads of the characters' stories together, the camera moves through the gathering at Méliès' home, the choreography integrating the camera and actors' movements in time and space. In the very.co.uk adverts, the camera and actors' movements in time and space are disconnected through the mediating influences of technology. As the camera arcs through the spaces of the motionless human figures, there is a sense of coming upon these spaces and figures differently, their stasis striking in the context of the mobility of the camera. The very.co.uk advert series play on an altered configuration of space offering a distinctive experience whose geometry of movement makes explicit its mediation by digital technologies. Even though the space in which actions are taking place remain configured as three-dimensional space based in physical reality, the motion perspective adds a set of contours that map onto a space with an alternative range of parameters. Movements created using 3-D animation software too have the capacity to map onto another reality, that of a digital space, generating what can be called digital contours.

To say more about digital contours, the following develops the idea of 'more-than-representational' space. By taking into account the relations between things, it creates an opportunity for describing the

DOI: 10.1057/9781137448859.0006

transformations a mediating technology can make to the experience of viewing images. To develop this account several ideas are drawn together. The first is digital space has the potential, which is not always exploited, to offer a distinct kind of movement in relation to virtually projected objects. The second is this movement can be linked to 'software materiality', an idea developed more fully through an approach informed by software studies and through a return to interviews with software users. Thirdly, it remains the case that a viewer experiences space as a mixture of representational and more-than-representational space, which in narrative cinema and animations is often based on both perceptual realism and digital spaces. How these organizations combine is the basis of a different kind of viewing experience.

The difference digital makes

Many discussions about visual effects, computer-generated animation and games address the kinds of images constructed using software in terms of what they look like, from the photoreal simulations of humans, environments and other entities, to the different stylistic possibilities of visual effects filmmaking, games and animation (Telotte 2010; Kirkpatrick, 2011; Purse, 2013; Whissel, 2014). Representational approaches have a long tradition in moving image studies and continue to provide a significant arena for debate. Whether talking about cinema, television, games or animations, moving images are depictions of actions. These depictions have a complex relationship with the world and variously dislocate, articulate and translate actions between the social and political worlds and that which is depicted audio-visually. Moving image practices re-assemble and re-associate actions within the over-arching narrative of a fiction or non-fiction, game or animation or live-action film, and also the stylistic conventions of particular types of image making. Thinking about software in the creation of a digital space need not only refer to what is seen. It can include paying attention to encounters and experiences mediated by movements whose origins are digital as opposed to physical. *Rango* is an animated feature drawing on storytelling conventions from the western genre and its iconography, and there is chase sequence between the hero chameleon Rango and the Hawk, the latter frequently terrorizes the inhabitants of Dirt. Shown as a seamless sequence, the virtual camera follows the chase along the spaces

DOI: 10.1057/9781137448859.0006

between houses, moving from low to high, and in and out of proximity to the chasing figures. The shifting proximities and rapid movements create a distinct experience of the entities being represented. Their actions, their turn about, and turn about again chasing as they smash through wooden fencing throwing debris into the air, can be described both representationally and also more-than-representationally. The latter speaks to the torsional experience of speeds and proximities, leading to a different way of thinking about how space turns up in a digitally mediated experience of reality.

The idea of more-than-representational space draws on non-representational theories found in the work of Nigel Thrift and other cultural geographers including Hayden Lorimer and Sarah Whatmore. Non-representational approaches pay attention to the relations between bodies (both human and non-human) and their environments. The central question is to uncover how sense and significance emerge from on-going practical action, rather than stepping back and contemplating such actions as events 'drained for the sake of orders, mechanisms, structures and processes' (Dewsbury et al., 2002, p. 438). Since much non-representational theory is directly concerned with making sense of actions as they happen in the world, its bearing on images is not necessarily obvious. Images are always enmeshed in structures and processes, through the technologies of their construction, the audio-visual genres they work within and the thinking of artists involved in their creation. Thrift's discussion of software and space and movement-space nevertheless provides insights that can be effective in thinking about moving images. Writing about the ways in which software are increasingly embedded in the world, he argues that the background time-spaces of people's experiences are 'changing their character, producing novel kinds of behaviours that would not have been possible before' (Thrift, 2007, p. 90). This modified space creates the potential for novel kinds of behaviours. In their work on coded spaces, Rob Kitchen and Martin Dodge draw on Thrift when they argue: 'Software matters because it alters the conditions through which society, space and time, and thus spatiality are produced' (Kitchen and Dodge, 2011, p. 13). Though talking about software's influence on physical space, their point is also valuable when thinking about 3-D animation software. The digital spaces created using software are similarly changed in character. The possibility spaces of software alter the opportunities for creating space. And as well as enabling the mimicry of physical reality, it also has the potential allow

DOI: 10.1057/9781137448859.0006

the generation of novel kinds of behaviours in space. *Wreck-It Ralph* (USA, 2012) illustrates this nicely when the story-world shifts from simulated 8-bit space to 3-D animation. At first, the character Ralph is introduced in the context of the arcade game of which he is a part. The game is played on the flat façade of an apartment building, with Ralph and other characters simply moving on the surface. Movements remain two-dimensional, up, down, left and right, with no movements along the Z-axis or into depth. In keeping with the perspective of playing an arcade game, viewers see Ralph from an orthographic perspective: no depth projection and the relationship between objects in the screen space remain constant. After the virtual camera passes through the screen of the arcade game console and into the alternative reality behind the screen, the characters are depicted in 3-D perspective. *Wreck-It Ralph* introduces this 3-D world from the perspective of an ostentatious virtual camera twirl around the same apartment building, now modelled in 3-D and depicted using depth projection. The seamless circling movement from top floor to ground level accentuates the difference of 3-D space. Since the opening scenes of *Wreck-It Ralph* have taken place in the simulation of a 2-D 8-bit arcade game, the transitional moment serves the purpose of displaying the alternative representational possibilities of 3-D space. But the movement through this 3-D space conjures something more than a depiction of representational space. In the contrast between the flat dimensions of the 8-bit arcade game and the 3-D space behind the scenes of that game, there is sense of coming upon space differently. This sense of coming upon space differently occurs in a move between representational and more-than-representational spaces. In representational space, objects and entities are depicted; in more-than-representational space, virtual camera movements provide the experience of 'a different sense of how things turn up' (Thrift, 2007, p. 102).[16]

Knowing more about the digital contours generating these more-than-representational dimensions includes having a keener understanding of the software used in creating moving images. As we saw in the previous chapter, when working with software, users enter into a complex relation involving layers of contexts and practices. Their creative work is a folding together of influences from their accumulated experience, the particular project and its situation, and also any software that they are using. In this account, software is active in the creation of objects and entities. The following continues thinking about digital contours through a more specific focus on Autodesk Maya. It again draws on paratexts and

DOI: 10.1057/9781137448859.0006

interviews with users of the software, using these to explore the digital contours of 3-D animation. When encountered via the software's UI, the modelling and animation toolsets of Maya are accessible via drop-down menus. The UI also includes a viewport where modelled objects are virtually projected. Within this viewport, by activating the manipulator tools, users move through the virtually projected space. In the previous chapter, I described the viewport as virtually projecting a space equivalent to three-dimensional space. My focus here is on the experience of working in that space. When using manipulator tools to engage with objects in virtually projected spaces, even though dimensionally similar to those of physical reality, the objects have a different degree of accessibility than those in physical space. Thinking more about this different quality of accessibility provides a way of exploring how the mediating influences of software generate contours in a digital space.

Animators and modellers talk about the mediations of software. Martyn Gutteridge comments on the 3-D environment of Maya through a contrast with his experience of hand-drawn animation:

> ... it is a bit simpler when you're doing it initially in drawn because you are completely in charge of the illusion. Whereas in the computer you are not completely in charge of the illusion, a lot of that is being created for you and you are manipulating that space.

Gutteridge's remark draws attention to the hybrid quality of 3-D space created during the process of using an animation package: when modelling, a user fills or creates a space, but at the same time the computer creates space too through the execution of numerous algorithms. Asked whether he thought there was something distinctive about 3-D space, game designer Jacky Jiang of thatgamecompany replied by reaching towards the idea that digital entities have a particular quality: '... when working with the medium you get familiar with it, you really see the physicality, I mean digi-cality because it is not physical ... You really get to know the quality of what it can produce ...'

A place to begin looking at what 'digi-cality' might entail is the modelling toolsets on the UI of Maya. Finding digi-cality is not always straight forward, as the mediations of software can appear to be hidden. When modellers comment, for instance, that working with 3-D animation software is like working with physical models, meaning models constructed from plasticine, balsa wood or other materials, the explicit influence of software recedes when hidden behind these physical parallels. Animator

DOI: 10.1057/9781137448859.0006

James Farrington comments: 'One of the things that struck me, working in Maya is a lot more like model animation than it is like drawn animation.' An experienced generalist, D, working in a large production house, also remarks:

> The path I took into it was a lot less intuitive, and there was a lot more maths. I mean this is a lot like Lego building on the screen because things are more advanced now. It's kind of done in front of you. Whereas I don't think I would have been able to cope in the past where it was a case of creating a cube with these dimensions, and typing.

In making an analogy between Lego building and modelling in 3-D space, D draws attention to the representational logic of building a model. When creating a model, the user chooses which set of primitive shapes to work with (NURBS, polygons or subdivisions). Primitive shapes are the building blocks used to initially generate the basic shape of the model. The building involves adding shapes to one another to make a different shape. Engaging with this kind of space relies on representational thinking, in the sense that it is focused on making something take shape. Asked if working in 3-D computer-generated space requires a different way of thinking to real world 3-D space, Aaron Chan comments that is similar but the materiality is different:

> I think it's very similar, and I think it's kind of great because we're not bound by materials or by heavy manual labour, but you can really create whatever you want to create. I think that is one of the main reasons why I started doing it in the first place, to be able to create worlds and bring things to life. That's the best part about 3-D, absolutely.

Chan's point about the different materiality of working with software as opposed to physical entities, need not only be limited to thinking about weight or labour, but also the affordances offered by software and what is added as opposed to being left behind.

In addition to all the possibilities of the modelling and animation toolsets, what is added by software is an enhanced capacity to move around and manipulate models. Manipulator tools, which can be controlled using keyboard and mouse combinations, facilitate three kinds of movement: rotate, pan and zoom. When working with these manipulator controls, users have full access to the virtually projected objects. The enhanced access is less obviously linked to the protocols of particular toolsets, and more to the ability to move within and around the projected space of

DOI: 10.1057/9781137448859.0006

the object. Paying attention to access given by the manipulator tools to objects makes more-than-representational experience and novel behaviours of digital space more evident. The degree of access, its potential for proximity and distance, the ability to move in and around objects, afford a distinct kind of engagement based in digital as opposed to physical reality. Commenting on moving through the space surrounding models, Paul Hilton of Arts University Bournemouth explains:

> I think when you're modelling something, you're visualizing the object in three dimensions. You are moving it around constantly, sort of working on, almost the back of it. It's almost like it exists in front of you. And you are aware of what you're doing at the front, and how that might influence the back of the object.

Hilton's description of modelling gestures to the way space around a model becomes active in 3-D. Looking through the viewport, users of Maya see a virtual projection of a digital object. The viewport mediates by giving a modeller the impression of getting close to the object, placed in greater proximity and able to manipulate from all sorts of orientations, at greater levels of detail and range of scales. When in perspective view, the viewport controls allow a modeller to rotate their digital objects, to see it from all sides, while also zooming into and out of the space to get closer to the detail of the model.

Figure 2.3 shows a set of stills of the same object seen from different positions. Together they illustrate some of ways modellers move within the viewport. When seen in action, modellers often quickly move within the viewport space, rapidly shifting around as they manipulate mesh structures at various degrees of proximity. Figure 2.3.a shows the brown hessian sack seen in previous figures in mid-distance; Figure 2.3.b demonstrates a closer view of the sack, with the weave of the hessian more visible. Figures 2.3.c and 2.3.d show the sack from underneath and at distance, respectively.

The space in which the model exists is digital and so not quite as we know it. The algorithms activated by the manipulators allow a user to experience movements around the model based on contours generated in a digital environment. Expanding further on the idea of contours generated in a digital environment, movements have the appearance of occurring in the full 360° of a projected space. This includes around the underside of the model, as the visible grid seen in the viewport is only a place marker and not a surface (unless defined as such). Accessibility

FIGURE 2.3   *Range of movements available with Autodesk Maya's viewport*

Note: Screen shots showing four orientations around a brown sack to illustrate the accessibility of virtually projected shapes in Autodesk Maya's viewport. See main text for further explanation.

Source: Images ©Paul Hilton.

DOI: 10.1057/9781137448859.0006

to the space is different from that based in human physicality, as are the qualities of movement. It could be argued for instance, that someone building a model in physical reality is able to pick it up and see it from all sides. But they remain unable to zoom in and out at pixel level resolution, to turn around the object instead of turning the object around. Given the high degree of accessibility and degrees of movement, the manipulator controls offer a perspective that is marked by its digital origin. Such traces of a technological origin in moving images have already been noted in relation to camera movements and visual effects in the cinema, and recently William Brown argued that digital cinema takes us beyond 'human perspective' (Brown, 2013, p. 48). Describing how the post-human is visible in digital images, Brown comments: 'Not only is digital cinema full of inhuman characters performing impossible feats, but it is also full of impossible camera movements and perspectives that seem to take us beyond the frame as we typically understood it' (ibid.). These impossible camera movements and perspectives take viewers to spaces usually inaccessible to a human perspective. Brown develops his position that such imagery, especially when mapped onto curved time and space conceptualized in contemporary physics, can be considered as an example of realism:

> It is paradoxical that in digital cinema these spaces are constructed through the use of the discrete functionings of the computer, which typically constructs space according to fixed, Cartesian coordinates. Nonetheless, the malleability of space in digital cinema, a democratization of space that allows us not to privilege certain points over others, takes on an unexpected level of realism when viewed from the perspective of contemporary physics. (ibid., p. 50)

The post-human perspective generated by the malleability of digital space is linked in Brown's account to a version of physical reality defined by contemporary physics. Even as this way of thinking addresses the malleability of digital space, it does so in a way that relies on parameters outside of digital space, or the functional space of a software used to the create the images and movements that appear on a screen. Approaching the parameters of space through the manipulators of a viewport, instead links digital spaces to a non-human perspective, with the different sense of how things turn up defined in terms of being 'more-than-human'. This phrase is borrowed from cultural geographer Sarah Whatmore who describes more-than-human as the ways in which non-human elements

DOI: 10.1057/9781137448859.0006

are part of the assembly or co-fabrication of socio-material worlds (Whatmore, 2006, p. 603).

Digital contours experienced via the manipulator tools delineate some of the parameters of a more-than-human digital space, and a brief detour into Matthew Kirschenbaum's work on computation and materiality provides a backdrop for thinking more deeply about the degree of non-humanness associated with this more-than-human digital space. Writing about the necessity to go beyond (or perhaps behind) the screen, Kirschenbaum suggests: 'Screen essentialism becomes a logical consequence of a medial ideology that shuns the inscriptive act' (2008, p. 43). The medial ideology to which Kirschenbaum refers is where digital writing is held apart from the inscriptive acts through which it is generated, aligning such writing with the instant transformation of visible screen-based events such as flicker, evanescence and lines of light. Such visible screen-based events are to be contrasted with inscriptive acts, examples of electronic textuality better understood via an account of software materiality. For Kirschenbaum, there are two kinds of such materiality: forensic and formal. Forensic materiality refers to individual data comprising a digital entity defined and located in time and space, such as saved data on a drive of some kind. Formal materiality, which is relevant to this discussion, refers to an instance created through a specific procedural operation. Which is to say, the output of a successful execution of an algorithm that follows the input of a particular set of data, as opposed to an abstract line of code. As Kirschenbaum puts it, formal materiality is: 'one where any material particulars are arbitrary and independent of the underlying computational environment and are instead solely the function of the imposition of a specific formal regimen on a given set of data and the resulting contrast to any other available alternative regimens' (2008, p. 19). Formal regimens, or algorithms, are one of the things that define the behaviour of digital things. This includes the algorithms behind the manipulator tools giving rise to movements and proximities in digital space. Accordingly, digital contours arise out of the materiality of software. In the previous chapter, I argued that user and software folded together, and where Kirschenbaum's emphasis on algorithms tends to defer the presence of the human users of the software, N. Katherine Hayles discussion of materiality is useful in providing a reminder of its hybridity: 'Materiality is unlike physicality in being an emergent property. It cannot be specified in advance, as though it existed ontologically as a discrete entity. Requiring acts of human attentive focus

DOI: 10.1057/9781137448859.0006

on physical properties, materiality is a human-technological hybrid' (Hayles, 2012, p. 91). The same is true of creating and working with space when using Maya. The digital contours visible on the screen emerge through the interaction of a user with the functional properties of a formal regimen. In this sense, digital contours offer a more-than-human perspective rather than one that is wholly non-human. A more-than-human perspective generates more-than-representational space.

Through Kirschenbaum and Hayles, I have described the ways non-human elements are part of the assembly or co-fabrication of the more-than-human material worlds created using software. The possibility space visible in the viewport of Maya includes the full range of movement achievable in digital space, and such a range of movements is shared with other 3-D animation software packages. Tumbling movements doable with rotate tools, extreme proximities or scales of distance available in zoom, or unrestricted panning movements combine to make space accessible. The absence of gravity and weight, in a physical sense, contribute further to the non-human elements. Even so, movement within a data heavy scene would be slow because of the amount of computation necessary to enable the manipulators to relocate in real-time. This latter fact is a reminder that when moving around the viewport, a user is not moving around a space that exists, but is moving around a virtual projection of space re-calculated with each and every change. The limit is that of the speed and capacity for computation not physical materiality.

The intervention of software in the creation of a digital space refers not only to what is seen but also to how spaces are seen, encountered and experienced through software-mediated movements in digital space. The mediations of software are involved in the co-fabrication of the representational as well as the more-than-representational space in the viewport. The viewport is, however, the working space of Maya users and the images are seen and experienced in ways different to those rendered for screens. Only when playing a 3-D animation game is it possible to get close to the degree of access to space found in the UI of Maya or similar kinds of software. By contrast, in the context of visual effects, animation and also data visualizations, digital space is filtered through various kinds of shot compositions, whether those matched to live-action, the particular reality of a story-world or the accuracy of a data visualization. Speaking of his experience of learning to use Maya, Ben Thomas, then a layout artist with two years' experience of working in the industry, talks of curbing the fun of using virtual moving cameras through effective shot composition:

> Yes the teacher at the time, [name removed], he said lots of people when they start Maya tend to just tumble around in the 3-D world, because it is really, really good. You can put the camera anywhere… And he always used to go 'tumble tumble render'. … With that comment he was basically saying compose your shots. Don't forget the kind of knowledge you've brought with you, and that thought process. Don't get caught up in just having the ability to go anywhere. Still think about what's the purpose of the shot, what you're trying to say, what you're trying to achieve.

The tutor's reminder about being mindful of shot composition attends to a conflict between the two sets of potentialities of representational and more-than-representational spaces. Conventions of shot composition temper the ability to go anywhere, privileging instead the pursuit of an organization that draws on existing aesthetic and storytelling devices. As Ben Wiggs, an animator working Double Negative, comments:

> For animators, we are probably the most two-dimensional part of it. To make an animation you are making a movement that looks nice on a screen. So we really are creating what's occurring, what the audience is going to see. So very often, I'll be making something move just so long as it looks right in the camera view. Something will create a nice arc, a sweep of movement… but actually if you twirled the camera and saw it from a different angle, you'd think that looks all wrong. Because it's not necessarily physically accurate but if it looks nice from the camera, and the director's happy, then…

Wiggs' remarks gesture to the more open possibilities of movement and perspective in the viewport, which are often constrained. Working to a particular camera view limits the animator to a specific shot configurations.

The possibility spaces of modelling and animation tools are more open than the spaces of visual effects, animations, or data visualizations. In the finally screened images, digital contours are filtered through particular representational conventions and the requirements of storytelling or knowledge sharing in a given set of moving images. As a consequence, images contain both representational and more-than-representational spaces, and the ways they co-exist is considered in a discussion of examples drawn from a range of contexts: the trailer for *World of Warcraft: Mists of Pandaria* (USA, 2012), the live-action film *Oblivion* (USA, 2013) which relies on many visual effects sequences, the animated feature

DOI: 10.1057/9781137448859.0006

Rango, data visualizations *Molecular Visualizations of DNA* (Australia 2003), and *Hollow Bjork*, and the game *Journey* (USA, 2012).

Representational and more-than-representational spaces

Computer-generated images are aesthetic creations, representational images including imagined spaces and visualizations of molecules, playable spaces, or products of a particular set of working circumstances. Such interpretive frames enable complex and sophisticated perspectives, although the discussion finally remains on familiar territory. Photorealism and perceptual realism, for instance, draw digital entities into a world defined through physical reality. As a consequence, the difference software makes remains active only in support of these interpretive frames. Drawing on non-representational theory offers a way of thinking beyond representational paradigms, and to talk about more-than-representational spaces. In examples discussed earlier, the virtually reconstructed crime scene in *Iron Man 3* and the very.co.uk advert series for their Definitive line of clothing, space is not only modelled differently because of digital technologies, but also reveals its mediation by digital technologies. The modelled or modified spaces where action occurs map onto a three-dimensional space based in physical reality, while motion perspectives add a set of contours which map onto digital space.

Digital contours, when they feature strong degrees of proximity and movement, add another dimension to our experience of computer-generated images. This experience is an affective one, created by the torsions of an unusual degree of proximity or turn around something in digital space. It is important to note here that more-than-representational space co-exists with representational space. Unusual proximities and arching untethered movements are perceptible because they are close to or turn around something. The distinctiveness created by such proximity and movement lie in what they add to an encounter with objects and entities otherwise familiar. The geometry of the object goes 'out of phase' in this encounter; it looks familiar, but the motion perspectives of our approach are unfamiliar, digital in character. When writing about a participant's experience of a virtual space installation, Mark Hansen noted: 'Space...is constructed via the affective connotations provoked by the participant's journey through the virtual space'

DOI: 10.1057/9781137448859.0006

(Hansen, 2006, p. 182). The spaces of films and animation, visualizations and adverts, and even games, offer a different set of parameters to virtual space. In combining representational and more-than-representational space, they integrate the affective engagement engendered by digital contours with an engagement based on cognitively making sense of the space.[17] The question is not simply about what a space looks like, but how one looks at and comes upon it. Returning to my example of the transition between 2-D and 3-D space in *Wreck-It Ralph*, the building is constructed in 3-D, but what it looks like is less interesting than the virtual camera spiralling around and taking viewers through that space. Nigel Thrift suggests: 'spacetime arises out of multiple encounters, which though structured, do not have to add up: as myriad adjustments and improvisations are made, so new lines of flight can emerge. The fabric of space is open-ended rather than enclosing' (Thrift, 2007, p. 98). If spatial engagement involves a multiplicity of encounters, then encounters with moving imagery not only concerns aesthetics, narrative or performance, but also a sense of coming upon something. By keeping in mind the notion of lines of flight, and allowing digital contours to not be determined by familiar frames of reference, they remain as more-than-representational moments based on an affective encounter generated as movement or proximity.

Before commenting on specific examples of more-than-representational space, I want to briefly note one of the difficulties of looking for traces of digital contours: movements potentially offering a more-than-human perspective might be fully contained within familiar frames of reference. In animations created for compositing within a live-action sequence, any movements tend to be carefully created to match those of the live-action shots, a point illuminated by James Farrington:

> I've worked on shots where you might have a forest like in *Harry Potter*, spiders coming down the tree and the forest is CG, and it all looks very realistic, and the spiders are CG, and then the camera pans down and there's two live action characters in it, but they're the only live-action elements. And then other times, like the Hippogriff for instance, there will be a creature that just sits in a live action setting. And, I think, when you're doing that style of animation, the demands on your animation are quite realistic, and very exacting.

In animated sequences embedded in live-action film, the configurations of images and motion perspective are linked to those given in the story-world, which only rarely completely step outside conventions of

DOI: 10.1057/9781137448859.0006

physical reality. In animations where the movements or the physicality of live-action need not be so closely matched, there is greater potential for the traces of digital contours to be visible. The cinematic trailer for Blizzard Entertainment's *World of Warcraft: Mists of Pandaria* shows two facets of 3-D animation software. The first can be linked to representational frames through the physicality of the figures. This is a point addressed by Aaron Chan when speaking about the possibilities of being able to push beyond live-action simulation: 'You know, we do make photoreal humans, but even those humans are not fully real. We might pull some of their proportions or make sure that their faces are longer than normal. There is usually an extra something.' Such pulled proportions are especially evident in the human who battles with the orc at the opening of the trailer. The figures are quite matched in terms of their respective shapes, but the man's chest and arms are beyond the proportions of a human body. Against such expanded proportions, the man's head appears relatively small, even though the square jawed hero look of his chin and cheekbones are more than usually pronounced. The second facet of 3-D animation software emerges in the proximity of the (virtual) camera placements. A shot begins with a distant view of the wreckage of a ship in the shallow water of a bay, with the bottom of the frame trimmed by the edge a rock ledge, shown just out of focus. Against the musical background, a couple of explosions punctuate the soundscape, and just as a grunt of physical exertion is heard, the focus shifts to reveal both the textures of a log and the green skinned hand of an orc. This hand is in extreme close-up, filling almost a third of the screen, while the background view is quickly taken up with the upper torso of the orc who has climbed the rock face. Pulling himself over the edge, the orc suddenly launches his body into a space where a live-action camera would likely be placed. The figure's movement into that space, though brief, is unexpected. The cut to the next scene, which is preceded by a black empty screen, opens with another configuration of space in which the relationships between the parts and the whole are pushed. The camera is set low and directed upwards, taking in the end of a stake in full focus, with the man sharpening it out of focus. Again, the sound design here supports the visible action: the sound of a sharpening blade against wood. The music carries just a hint of tension, perhaps based on the expectation that the two figures will soon meet and most likely fight. As the focus pivots towards the man, leaving the stake blurred on its edges, the configuration of space has a torsional quality. The low

DOI: 10.1057/9781137448859.0006

angle, tightly slanting upwards and in close proximity to the stake, has a feeling of being forced. Looked at in terms of a representational framing only, the detailed textures of the all of the elements within the frame are impressive, and the pulled proportions of the figures effective in showing brawn rather than brain. But the more-than-representational space given in the relations between all of the elements of the scene are more unexpected. They add a tension to the imagery, which can be attributed to an anticipation of potential action. At the same time, it reveals the ways in which digital mediations reconfigure an experience of representational spaces as more-than-representational, in this instance through an emphasis on pushed proximity.

Oblivion (2013) is a science fiction film that too combines representational and more-representational spaces, in this instance the movements mapping onto the non-human entity of the film. Though this has the potential to close the gap between representational depiction and more-than-representational experience, the movements of the non-human entity still offer viewers a different sense of how things turn up. One such sequence is when drones destroy the Scav base camp at Raven's Rock. The Scavs, it is revealed through the narrative, are the few remaining humans on earth battling the aliens who have invaded earth using clones and drones. The drones arrive at the base camp in search of Jack Harper, the last drone repairman on earth. The latter has been recruited back to the Scav resistance at the point of the drone attack. At the beginning of the skirmish, the drones invade the hangar space and do battle with the Scav forces. In this opening sequence, rapid camera movements are tied to the same physical ground as the humans, whether tracking Jack as he falls to the ground, moving towards him, or framing other Scav fighters. Mid sequence there are some 500 computer-generated shots (Desowitz, 2013). Released from the same physical constraints as the human, the virtual camera plays a kind of tag with digitally constructed drones: arcing through the spaces, falling and rising vertically, moving towards, or hanging back. In the swift, unrestricted and almost swinging mobility of this brief sequence, space feels different, capturing perhaps the logic of the momentary ascendency of destructive drones. Indeed, the sequence ends back on human ground, where the drones are eventually destroyed.

It is interesting to contrast *Oblivion* with *Gravity*. *Gravity* relies on many animated and visual effects sequences to generate the visual dimensions of space and also to create the action sequences that make up much of

DOI: 10.1057/9781137448859.0006

the film. Even though a number of scenes in *Gravity* feature characters, especially the central surviving character Ryan Stone, tumbling around in space, the more-than-representational dimensions of digital contours are not strongly present. *Gravity* opens with a black screen on which appear statements conveying the hostility of space 600km above the earth's surface, which is where the action of the film takes place. Once the film's title appears, the rising score abruptly stops on a cut to a partial view of earth filling four fifths of the screen, a slow silent turning view of the planet rotating into 3-D depth. As the stunning blue of the ocean and swirling masses of cloud give way to a landmass, a small speck in space gradually gets larger and radio transmissions become audible. The small speck turns out to be an American space shuttle whose team is fixing the Hubble telescope. The pacing of the scene quickly shifts to introduce the life-threatening disaster of a rapidly approaching shower of space debris, which kills all the crew bar two, Ryan Stone and Mike Kowalski, and destroys the space shuttle. Throughout, the filmmakers have been at pains to realistically recreate the microgravity experienced by astronauts 600km above earth, and so movement and motion perspectives are aligned with that aim. In the opening 'shot', a 17-minute seamless sequence of images, the camera, though very mobile, remains connected to the microgravity of the characters as they move around the space shuttle. While there are brief moments when the camera seems to be untethered from this perspective, mostly the constant reorientation of the re-framings follows the movements of the characters.

Looking behind the scenes of *Gravity* reveals the way in which the camera was suspended in a rig, and a complex choreography between actor and camera was developed to achieve the mobility of the shots. For instance, the breakdown of the shot where Stone enters the International Space Station (ISS) shows the extent to which the actors and filmmakers worked to create the illusion of weightlessness. In this sequence, Sandra Bullock (who plays Ryan Stone) was seated on a bicycle seat rig, with her legs held in place so she was free to move her upper body. Her movements, slowed to a third of normal speed, were learnt like dance steps to hit various marks. But rather than the marks being tape on a floor, these were more labile. As Bullock performed the slowed-down actions of removing the gloves, helmet, leg and chest panels of a space suit, three elements rotate around her: the camera on a computer arm, a light behind a partial set to give the impression of the very bright sunlight of space coming in through the porthole of the ISS airlock, and the image

DOI: 10.1057/9781137448859.0006

of earth seen through the same porthole. These three elements rotate around each other, and Bullock's marks are particular alignments of the three elements. The details of how *Gravity* was designed and shot reveal the actors and filmmakers to be working with a spatio-temporal logic informed by microgravity. Like many visual effects sequences, effort is taken to mimic this physical reality and though a highly mediated process, the traces of technological mediation are much more hidden. By contrast, in *Oblivion* movements through space are untethered from a spatio-temporal logic based on physical reality. Even though the action of movement occurs in a realistically rendered and constructed space, the movements of the camera are mediated beyond the expectations of physically grounded movements. They operate according to a logic that maps onto the more expanded parameters of digital contours.

The animated feature *Rango*, Industrial Light and Magic's (ILM) first foray into animation, reveals digital contours too, though not ones that have the potential to so literally map more-than-human movements onto non-human mechanical entities considered in *Oblivion*. The computer-generated imagery of the animated western is rendered in extraordinary detail, visible in the textures of the creatures such as the chameleon Rango, the hero of the animation, or Bad Bill, the gila monster leader of the outlaw gang. In a publicity and marketing interview for both *Rango* and Autodesk, lead animator Kevin Martel comments:

> This is such a detail-rich world, we needed software that would enable us to see and show as much of that detail as possible as we were animating. We needed to see all of those wrinkles, bumps, ridges, hairs, and scales as we went along, so we'd know what it was going to look like on the big screen. Maya was simply spectacular at letting us do all of that. (Autodesk, 2011)

Though modified to fit into a cartoon reality, this kind of detail fits with the history of photorealism in ILM, or as Julie Turnock puts it in relation to the visual effects of ILM's live-action films: 'digital imaging ... replicates an accepted aesthetic *photo*realistically' (Turnock, 2012, p. 158; emphasis in original). Backing off from the detail and paying attention instead to the pushed proximity between the virtual camera and these figures, digital traces become apparent. For instance, during Rango's first meeting with Bad Bill and his gang inside the saloon of Dirt, there is a close-up of the villainous cat in which the camera seems almost to sit within the brim of its hat. Not just equivalent to an extreme zoom, it is also the trace of a digital contour where the scale of proximity can be so much closer. Following

DOI: 10.1057/9781137448859.0006

on from this first encounter with the gang, the action moves into the street, at which point there are two moments of very low and mobile camera placements, under Bad Bill and Rango's respective tails. In keeping with much of *Rango*, these shots are encapsulated within a series of comic and generic references, with the sound effects and dialogue grounding the images in the western-based story-world of *Rango*. But they are also examples of a virtual camera whose contours of movement activate the space of scene, bringing the usually uninteresting space between a figure's legs into the action. What might be called a pushed pose comes through in the proximities of camera and figure. Where pushed poses in traditional hand-drawn and computer-generated animation more usually refer to the pose of a figure, what seems to be pushed in these examples is proximity. There is something more extreme about how close things get. These examples of pushed proximity generate a different sense of spatiality, even as it is held in place by the humour and the generic devices of the story-world sound and images. As viewers, we are left with the sense that a very familiar space is active in different ways, and digital technologies have the capacity to alter the terms of proximity in the relations between things.

Proximity is also central to the imagery of the molecular movies of Drew Berry. In designing his animations, Berry aims for balance between the visualization of molecular events and a dramatic tension drawing viewers into the world of the images. The more-than-representational experience of space in these images adds tension, especially in the first half, where proximity between objects and a virtual camera is exploited in the animation *Molecular Visualizations of DNA*.[18] The opening, which has a voice over narration as well as the sound design, frames the action from a static view. Even though the shots are mostly static, the molecules are in constant motion, threads of folding chromatin progressively filling the screen as they increase in size. The proximity of these moving threads gives an impression of being within their environment; a positioning that is enhanced by the sound design created by Franç Tétaz. Combining a machinic pulse with an echoey liquid ambient sound, along with the close framing on detail, the sound design elicits a strange proximity to events. The close proximity might conjure an interpretation based on realism linked to microscopic scale, but the more objective presentation of such imagery is tempered by motion and changing proximities, as well as the music and sound. The same proximity is evident in Berry's animation for *Hollow Bjork*, released as part of Bjork's *Biophilia* album

DOI: 10.1057/9781137448859.0006

(2011). The animation opens with a mobile shot passing through a nest of capillaries before settling in close proximity to a series of moving cellular elements: skin tissue, a cell in prophase and the mitotic spindle, before again becoming mobile and passing through cytoplasm.

Proximity, then, is one facet through which digital contours become more evident. But as I have noted, digital contours emerge through the arcing continuous motion seen in *Oblivion*, the frenetic chase between the Hawk and the chameleon in *Rango*, as well as movements that conjure a sense of distance, scale and perspective. Such kinds of movements are associated with games through unbroken first or third-person perspectives, and their presence in live-action cinema are interpreted as examples of remediation (Jones, 2007; Brooker, 2009). Lisa Purse and William Brown separately make arguments loosening the virtual camera from existing aesthetic strategies, arguing that movements occur through any axes and any scale within digital space (Purse, 2009, p. 225; Brown, 2013, pp. 42–43). Taking these ideas further through more-than-representational thinking, the award-winning *Journey* exploits the device to an unusual degree. Its gameplay is based around a figure seeking a higher plane of existence (literally a mountain in the distance with a light shining from it). The figure, once it has accumulated enough glowing icons, is able leap high into the air. As the gamer learns to move using these floating jumps, they navigate the terrain, often alone, though occasionally joined briefly by other figures. The movements available to a player in *Journey* do not have as full a range of digital contours as found in the viewport, but they go some way to eliciting a similar experience of space. This is not simply a feature of the availability of the movements, but also down to the thinking behind the game. As described by the game's designers, thatgamecompany, the gameplay, which is based on exploring a sparse sandy dessert via swooping and leaping movements, is an interactive parable through which to experience life's passage. Given the presence of digital contours, the passage of life experienced in *Journey* is a means of coming upon space in ways that are more-than-human: representational spaces with further contours mediated by digital technologies.

Conclusion

The difference software makes to the moving image has been the central thread of this discussion. Following the traces of its mediations started

DOI: 10.1057/9781137448859.0006

on familiar territory with the logic of realism often informing representational spaces of many digitally constructed images. In narrative
films such as *Guardians of the Galaxy*, *Thor 2* and *Iron Man 3*, this logic
predominates, as it does too in animations including *Despicable Me*, *The
Lego Movie*, games such as *The Last of Us* and data visualizations, with
the difference that software makes more often than not covered over.
Even so, it is possible to look beyond this logic and find moments where
images reveal organizations of time and space mapping onto different
kinds of realities, whether the imaginary cloud particles of *Thor 2* or the
digital space of *Iron Man 3*. As a way of deepening the thinking behind the
idea that software has the potential to reconfigure the relations between
things, my focus returned to ideas based on the organizations of software, and how it informs our understanding of configurations of time
and space in images on the screen. From the relative stasis of modularity
and the geometry of shapes, my discussion moved onto thinking about
the experience of digital objects as a combination of representational
and more-than-representational space. This experience relies not only
on how image elements are combined, but also on the ways in which
a viewer is given access to those images through the movements and
degrees of proximity created through digital contours.

The examples considered in *Oblivion*, the trailer for *Mists of Pandaria*,
Rango, and *Journey*, between them a live-action film using visual effects,
computer-generated animation and also a game, combine representational and more-than-representational spaces, with the latter emerging especially in pushed proximities and in movements of the virtual
camera. Representationally, the images of each of can be analysed in
terms of generic codes, uses of digital technologies to build novel spaces,
the styles of imagery. Non-representationally, the images can be analysed
in terms of traces of digital contours, and how movements and extreme
proximities add a sense of difference to the experience of the representational imagery, the basis of more-than-representational space. What
precipitates out in the co-existence of representational and more-than-
representational spaces is intangible, an affective dimension held in the
tensions of proximity, an unusual movement around objects as opposed
to the objects themselves. It is a quality of disparity with the potential to
cause viewers to reflect on the difference that digital technologies make
to the experience of time and space, and not just the distinctive shapes
they make of the world.

DOI: 10.1057/9781137448859.0006

Notes

1   William Schaffer describes 'the Pixar feel' of the *Toy Story* films as a 'poetic affinity between the plastic and the digital, so tangible in the character of Buzz Lightyear' (2004, p. 85).

2   *L'Odyssée de Cartier* was directed by Bruno Aveillan; filmed on numerous locations, the live-action footage of panthers was embellished using an extensive range of visual effects. The Moving Picture Company (MPC) created the *Ribena Berries* adverts.

3   Mark J. P. Wolf has written about the transmedial construction of imaginary worlds. His analysis remains focused on the transmedial opportunities for story-telling, rather than worlds with a distinct set of physical laws of reality (Wolf, 2012).

4   Mihaela Mihailova notes the sense of mastery evident in paratexts associated with constructing animated images (Mihailova, 2013). This sense of mastery can also be found more widely in moving image practices, described by John T. Caldwell in his analysis of trade stories in production culture disclosures (2008, pp. 37–69).

5   See, for instance, the several animations at http://www.mpa-garching.mpg. de/galform/data_vis/.

6   Annabelle Honess Roe (2013) discusses paratextual authentication operating in relation to *Walking with Dinosaurs* (UK, 1999) amongst other animated documentaries. The authority of space simulations similarly resides with a range of paratexts.

7   An alternative progression of games development can be found in online and app-based games for smart phones, where data weight is more restrictive in terms of real-time processing.

8   As has occurred for other games, fans of *The Last of Us* have edited together the cinematic sequences from the game, and uploaded them onto YouTube.

9   This explicit aim is explained by the team involved in making *The Last of Us* in the documentary *Grounded: The Making of The Last of Us* (USA, 2013).

10   I mentioned the scripting languages for Maya in the previous chapter, MEL and Python. Python is an object-oriented language, though MEL is not.

11   For a more detailed discussion of object-oriented programming see Casey Alt (2011) and Jim Keogh and Mario Giannini (2004).

12   Allan Cameron also uses the idea of modularity in relation storytelling (Cameron, 2008).

13   On its website, Animal Logic, the studio who created *The Lego Movie*, has an interview with various personnel involved in making the film. See http:// www.animallogic.com/Studios/Work/The-LEGO%C2%AE-Movie.

14   For its operators, using a Steadicam can be physically arduous. The rig is heavy and the camera operator has to move through a space, avoiding

DOI: 10.1057/9781137448859.0006

objects in their path, all the while making the subtle movements required for the sequence.

15   A breakdown of various moments of the music video can be seen at http://okgo.net/wotw/

16   Though this opening discussion has used the phrase non-representational theory (and space), the term is not without its critics. As it works well in the context of a discussion of what software brings to the moving image, Hayden Lorimer's suggested alternative of 'more-than-representational' will be used (2005, p. 83).

17   This is not to ignore the other affective dimensions of moving images, especially sound, only to say that digital contours add another dimension through proximities and movement.

18   http://www.wehi.edu.au/education/wehitv/molecular_visualisations_of_dna/

DOI: 10.1057/9781137448859.0006

Conclusion

Wood, Aylish. *Software, Animation and the Moving Image: What's in the Box?* Basingstoke: Palgrave Macmillan, 2015.
DOI: 10.1057/9781137448859.0007.

▶

Knowing more about software involves knowing not just what it does, but also how people use it, think about it as well as through it, when they construct digital things. As digital technologies continue to develop, so too does their potential to transform the ways we work and interact with the materials in which they are embedded. For animators working with software, it changes the ways they work with the entities they create, how they see and interact with them. Exploring what this means requires finding a way into software. As David Berry notes: 'What remains clear ... is that looking at computer code is difficult due to its ephemeral nature, the high technical skills required of the researcher and the lack of analytical or methodological tools available' (2011, p. 5). This high technical skill, and perhaps a perspective that technologies are 'only' tools, leads to a tendency to talk about digital things, including images, without paying attention to software: 'writings on digital media almost all ignore something crucial: the actual processes that make digital media work, the computational machines that make digital media possible' (Wardrip-Fruin, 2009, p. 3).

Software's opaqueness is counterbalanced when it is approached as a neighbourhood, ensuring its visibility as a multidimensional and shifting object. In *Software, Animation and the Moving Image*, I have put together a methodology that places Maya in the middle of things, a variety of materials including video tutorials, manuals, and interviews with users, as well as the wider historical developments of software. By taking this material as a neighbourhood, the software's UI is revealed as less an array of toolsets, and more a complex meshwork of humanly meaningful frames and patterns, of which toolsets are only a part. Using software is a complex entanglement between the software, its users, as well as its discursive and material histories.

In this entanglement, some kinds of relations have gained force, while others have not. Maya, for instance, like other similar high-end 3-D animation software, was developed for and is marketed to commercial sectors, whether for visual effects, animations, games or data visualizations. Consequently, software is mostly embedded in mainstream cultural practices, ones operating both locally and globally. Many of the moving images discussed here are games, films or animations associated with the USA or the UK, but the actual producers of sections of their digital imagery might be more globally dispersed. For instance, visual effect studios such as Rhythm and Hues and Moving Picture Company have offices in the US, Canada, and India, with the latter in the UK too.

DOI: 10.1057/9781137448859.0007

Similarly, between them ILM and Double Negative have offices in the USA, the UK, Canada, Singapore and Australia. Since moving images created in this sector predominantly aim for a logic of realism, whether as simulations of specific events or as representations of physical reality with greater or lesser degrees of photorealism, perceptual realism informs the representational space of many of these digitally constructed images. The digital contours I have been describing for Maya, organizations of time and space drawing contours from digital space, co-exist with these representational spaces. Digital contours reveals some of the ways in which the code of 3-D animation software operationalizes space, and though focused on Maya, these contours emerge from the parameters of 3-D software, rather than software more generally. Other kinds of software, whether Toon Boom, GoAnimate or the still relatively expensive Flash, will have their own contours and so offer a different set of insights into digital spaces.

Although I took Autodesk Maya as my particular case study, giving accounts of digital media informed through an understanding of their toolsets, digital contours, and formal materialities, has further potential for thinking about software more generally. It offers a way through the opaqueness frequently surrounding digital media in moving images and beyond. Software underlying digital media can be understood through a wider set of vocabularies than as a package that enables the execution of data understood as a flow of os and 1s. Entities, objects, and events come into being through engagements with software, and these have meaning in physical reality because of how they appear and function in the world. They also have patterns of organization that owe their shape and structure to the algorithms involved in giving them an appearance in the world. Since the organization of digital information is different to that of physical reality, new possibilities in the relations between objects and entities on the screen continually open up. Their traces are visible in the more-than-human mediations of technologies described as a co-existence of representational and more-than-representational spaces.

Beyond the immediate context of computer-generated images, coming upon space in ways that are more-than-human is an instance of the profound way in which the digital reconfigures time and space more widely. In 'the wired world', algorithms are executed in microseconds, immediate transactions or communications lessening the crossing of continents to a moment gone already, an action that both reduces and also expands time and space. Experiencing the digital in action, as ultra

DOI: 10.1057/9781137448859.0007

fast processing reconfiguring our experience of time and space by way of computer-generated images, are mostly thought about as differences in representational possibilities. Still, more-than-representational spaces are visible, and our attentiveness relies on being aware of the ways in which digital technologies transform relations in time and space.

DOI: 10.1057/9781137448859.0007

Bibliography

J. Abouaf (2000) 'Maya: "So Ya Wanna Be a Rock Star" Revisited', *IEEE Computer Graphics and Applications*, 20, no. 2, 7–11.

Alias|Wavefront (1998) *Learning Maya Version 1.0* (Toronto: Alias|Wavefront).

R. Allen (1983) 'Bionic Dancer', *Journal of Physical Education, Recreation and Dance*, 54, no. 9, 38–39.

C. Alt (2002) 'The Materialities of Maya: Making Sense of Object-Orientation', *Configurations*, 10, no. 3, 387–422.

C. Alt (2011) 'Objects of Our Affection: How Object Orientation Made Computers a Medium' in E. Huhtamo and J. Parikka (eds) *Media Archaeologies: Approaches, Applications, and Implications* (Berkeley: University of California Press) pp. 278–301.

B. Alves (2005) 'Digital Harmony of Sound and Light', *Computer Music Journal*, 29, no. 4, 45–54.

S. Andrews (2013) Closer to reality: photorealism in computer graphics, http://www.pcpro.co.uk/features/385879/closer-to-reality-photorealism-in-computer-graphics, date accessed 15 August 2014.

Autodesk (2008a) Autodesk Maya: core hungry and ready for action, http://images.autodesk.com/adsk/files/maya_solution_brief_final.pdf, date accessed 11 October 2014.

Autodesk (2008b) *The Art of Maya*, 4th edn (Indianapolis, IL: Sybex).

Autodesk (2011) Rango, http://area.autodesk.com/rango, date accessed 11 October 2014.

N. Badler, C. Philips and B. Webber (1992) *Simulating Humans: Computer Graphics, Animation and Control* (Oxford: Oxford University Press).

N. Badler and S. Smoliar (1979) 'Digital Representations of Human Movement', *Computing Surveys*, 11, no. 1, 19–38.

R. Baecker (1969) 'Picture Driven Animation', AFIPS: Conference Proceedings, 34, 273–288.

C. Bangert and J.C. Bangert (1975) Artist and computer, http://www. atariarchives.org/artist/sec5.php, date accessed 6 June 2014.

J. Belton (2008) 'Painting by the Numbers: The Digital Intermediate', *Film Quarterly*, 61, no. 3, 58–65.

D.M. Berry (2011) *The Philosophy of Software Code and Mediation in the Digital Age* (Basingstoke: Palgrave Macmillan).

I. Bogost (2006) *Unit Operations* (Cambridge, MA: The MIT Press).

I. Bogost (2007) *Persuasive Games: The Expressive Power of Video Games* (Cambridge, MA: The MIT Press).

W. Brooker (2009) 'Camera-Eye, CG-Eye: Videogames and the "Cinematic"', *Cinema Journal*, 48, no. 3, 122–128.

W. Brown (2013) *Supercinema: Film Philosophy for the Digital Age* (London: Berghahn Press)

M. Burns (2014) Why is this much-loved, revolutionary software being killed by its developer? http://www.techadvisor.co.uk/features/software/why-is-this-much-loved-software-being-killed-by-its-developer/, date accessed 10 October 2014.

N. Burtnyk and M. Wein (1971) 'Computer-Generated Key-Frame Animation', *SMPTE Motion Imaging Journal*, 80, no. 3, 149–153.

J.T. Caldwell (2008) *Production Culture: Industrial Reflexivity and Critical Practice in Film and Television* (Durham: Duke University Press).

T. Calvert (1986) 'Towards a Language for Human Movement', *Computers and the Humanities*, 20, no. 2, 35–43.

A. Cameron (2008) *Modular Narratives in Contemporary Cinema* (Basingstoke: Palgrave Macmillan).

J. Canemaker (2009) 'Winsor McCay' in M. Furniss (ed.) *Animation: Art and Industry* (New Barnet: John Libbey Publishers Ltd) pp. 95–104.

E. Catmull (1978) 'The Problems of Computer-Assisted Animation', *SIGGRAPH '78*, 12, no. 3, 348–353.

E. Catmull (2014) *Creativity Inc.: Overcoming the Unseen Forces that Stand in the Way of True Inspiration* (London: Transworld Publishers).

W.H.K. Chun (2011) *Programmed Visions: Software and Memory* (Cambridge, MA: The MIT Press).

DOI: 10.1057/9781137448859.0008

C. Csuri and J. Shaffer (1968) 'Art, Computers and Mathematics', *AFIPS: Conference Proceedings*, 33, 1293–1298.

D. Derakhshani (2012) *Introducing Autodesk Maya 2013* (Indianapolis: John Wiley and Sons).

Design-Engine Education (2011) Design-engine education: industrial & product design training in Pro/Engineer, Solidworks, Maya, Rhino, Alias, Adobe: a history lesson on Alias 3D Software, http://proetools.com/a-history-lesson-on-alias-3d-software/, date accessed 1 October 2014.

B. Desowitz (2013) A New Post-Apocalyptic Look for 'Oblivion', http://www.awn.com/vfxworld/new-post-apocalyptic-look-oblivion, date accessed 15 January 2014.

J. Dewsbury, P. Harrison, R. Mitch and J. Wylie (2002) 'Enacting Geographies', *Geoforum*, 33, no. 4, 437–440.

F. Dietrich (1986) 'Visual Intelligence: The First Decade of Computer Art (1965–1975)', *Leonardo*, 19, no. 1, 159–169.

R. Ebert (1995) *Toy Story* movie review & film summary, http://www.rogerebert.com/reviews/toy-story-1995, date accessed 20 September 2014.

T. Elsaesser (2008) 'Afterword: Digital Cinema and the Apparatus: Archaeologies, Epistemologies, Ontologies' in B. Bennett, M. Furstenau and A. Mackenzie (eds) *Cinema and Technology : Cultures, Theories, Practices* (Basingstoke: Palgrave Macmillan) pp. 226–240.

I. Failes (2013a) The dark side: behind the VFX of Thor: The Dark World | fxguide, http://www.fxguide.com/featured/the-dark-side-behind-the-vfx-of-thor-the-dark-world/, date accessed 11 August 2014.

I. Failes (2013b) Iron Man 3: more suits to play with, http://www.fxguide.com/featured/iron-man-3-more-suits-to-play-with/, date accessed 11 August 2014.

I. Failes (2014) Brick-by-brick: how Animal Logic crafted The LEGO Movie | fxguide, http://www.fxguide.com/featured/brick-by-brick-how-animal-logic-crafted-the-lego-movie/, date accessed 24 September 2014.

Film Animator Today (1971) The Film Animator Today: artists without a canvas, http://www.ieee.ca/millennium/computer_animation/animation_today.html, date accessed 16 May 2014.

C. Finch (2013) *The CG Story: Computer Generated Animation and Special Effects* (New York: The Monacelli Press).

DOI: 10.1057/9781137448859.0008

G. Fitzmaurice and B. Buxton (1998) 'Compatibility and Interaction Style in Computer Graphics', *Computer Graphics*, 32, no. 4, 64–69.

J. Fordham (2014) 'Thor: The Dark World: Valhalla Rising', *Cinefex*, 136, January, 1–31.

M. Fuller (2003) *Behind the Blip: Essays on the Culture of Software* (New York: Autonomedia).

M. Fuller (2008) 'Introduction, the Stuff of Software' in M. Fuller (ed.) *Software Studies/ a Lexicon* (Cambridge, MA: The MIT Press) pp. 1–13.

M. Fuller and A. Goffey (2012) *Evil Media* (Cambridge, MA: The MIT Press).

M. Fuller and A. Goffey (2014) 'The Unknown Objects of Object-Orientation' in P. Harvey, E.C. Casella, G. Evans, et al. (eds) *Objects and Materials: A Routledge Companion* (London: Routledge) pp. 218–227.

M. Furniss (2013) 'John Whitney's Path to IBM', *Animation Journal*, pp. 26–26.

A. Galloway (2012) *The Interface Effect* (Cambridge: Polity Press).

M. Girard and A.A. Maciejewski (1985) 'Computational Modelling for the Computer Animation of Legged Figures', *SIGGRAPH '85*, 19, no. 3, pp. 263–270.

M. Green (1991) 'Using Dynamics in Computer Animation: Control and Solution Issues' in N. Badler, B. Barksy and D. Zeltzer (eds) *Making Them Move: Mechanics, Control, and Animation of Articulated Figures* (San Mateo: Morgan Kaufmann Publishers) pp. 281–314.

C. Guly (1997) 'Ukrainian Canadian Computer Animator Wins Oscar for Technical Achievement', *The Ukrainian Weekly*, LXV, no. 4.

L. Gurevitch (2012) 'Computer Generated Animation as Product Design Engineered Culture, or Buzz Lightyear to the Sales Floor, to the Checkout and Beyond!', *Animation: An Interdisciplinary Journal*, 7, no. 2, 131–149.

M.B.N. Hansen (2006) *New Philosophy for New Media* (Cambridge, MA: The MIT Press).

N.K. Hayles (2012) *How We Think: Digital Media and Technogenesis* (Chicago: University of Chicago Press).

A. Honess Roe (2013) *Animated Documentary* (Basingstoke: Palgrave Macmillan).

E. Huhtamo and J. Parikka (2011) 'Introduction: An Archaeology of Media Archaeology' in E. Huhtamo and J. Parikka (eds) *Media Archaeology: Approaches, Applications, and Implications* (Berkeley: University of California Press).

DOI: 10.1057/9781137448859.0008

IPSoft (2014) *Amelia | IPsoft*, http://www.ipsoft.com/what-we-do/ amelia/, date accessed 9 October 2014.

M. Jones (2007) 'Vanishing Point: Spatial Composition and the Virtual Camera', *Animation: An Interdisciplinary Journal*, 2, no. 3, 225–243.

S.E. Jones and G.K. Thiruvathukal (2012) *Codename Revolution: The Nintendo Wii Platform (Platform Studies)* (Cambridge, MA: The MIT Press).

A. Kay (1996) 'The Early History of Smalltalk' in T. Bergin and G. Gibson (eds) *History of Programming Languages Volume 2* (New York: ACM Press).

J. Keogh and M. Giannini (2004) *OOP Demystified: A Self-Teaching Guide* (New York: McGraw-Hill/Osborne).

G. Kirkpatrick (2011) *Aesthetic Theory and the Video Game* (Manchester: Manchester University Press).

M. Kirschenbaum (2008) *Mechanisms: New Media and the Forensic Imagination* (Cambridge, MA: The MIT Press).

R. Kitchen and M. Dodge (2011) *Code/Space: Software and Everyday Life* (Cambridge, MA: The MIT Press).

K. Knowlton (1965) 'Computer-Produced Movies', *Science*, 150, November, 1116–1120.

N. Kurachi (2011) *The Magic of Computer Graphics* (Boca Raton: A K Peters/CRC Press).

J. Lasseter (1987) 'Principles of Traditional Animation Applied to 3D Computer Animation', *ACM SIGGRAPH Computer Graphics*, 21, no. 4, 35–44.

J. Linder, T. Price, C. Rosendahl and J. Lasseter (2009) 'Computers, New Technology and Animation' in M. Furniss (ed.) *Animation: Art and Industry* (New Barnet: John Libbey Publishing Ltd) pp. 199–206.

H. Lorimer (2005) 'Cultural Geography: The Busyness of Being "More-Than-Representational"', *Progress in Human Geography*, 29, no. 1, 83–94.

E. Luhta (2010) *How to Cheat in Maya 2010: Tools and Techniques for the Maya Animator* (Burlington: Focal Press).

A. Mackenzie (2006) *Cutting Code: Software and Sociality* (New York: Peter Lang).

N. Magnenat Thalmann and D. Thalmann (1996) 'Computer Animation in Future Technologies' in N. Magnenat Thalmann and D. Thalmann (eds) *Interactive Computer Animation* (Hemel Hempstead: Prentice Hall) pp. 1–9.

DOI: 10.1057/9781137448859.0008

L. Manovich (2001) *The Language of New Media* (Cambridge, MA: The MIT Press).

L. Manovich (2011) 'Inside Photoshop', *Computational Culture*, 1, 1–11.

L. Manovich (2013) *Software Takes Command* (London: Bloomsbury).

T. Masson (1999) *CG101: A Computer Graphics Industry Reference* (Indianapolis: New Riders Publishing).

V. Mayer, M.J. Banks and J.T. Caldwell (2009) 'Introduction: Production Studies: Roots and Routes' in V. Mayer, M.J. Banks and J.T. Caldwell (eds) *Production Studies: Cultural Studies of Media Industries* (London: Routledge) pp. 1–13.

S. McClean (2007) *Digital Storytelling: The Narrative Power of Visual Effects in Film* (Cambridge, MA: The MIT Press).

M. Mihailova (2013) 'The Mastery Machine: Digital Animation and Fantasies of Control', *Animation*, 8, no. 2, 131–148.

N. Montfort and I. Bogost (2009) *Racing the Beam: The Atari Video Computer System* (Cambridge, MA: The MIT Press).

L. Mori (2014) What's new in Maya, 3DS Max, Mudbox and Softimage 2015, http://www.3dartistonline.com/news/2014/03/whats-new-in-maya-3ds-max-mudbox-and-softimage-2015/, date accessed 29 April 2014.

A.M. Noll (1967) 'The Digital Computer as a Creative Medium', *IEEE Spectrum*, 4, no. 10, 89–95.

D. Norman (2013) *The Design of Everyday Things*, 2nd edn (Cambridge, MA: The MIT Press).

D. North (2008) *Performing Illusions: Cinema, Special Effects and the Virtual Actor* (London: Wallflower).

C. Pallant (2011) *Demystifying Disney: A History of Disney Feature Animation* (London: Continuum).

J. Parikka (2012) *What Is Media Archaeology?* (Cambridge: Polity Press).

F. Parkes (1986) 'Interactive Tools to Support Animation Tasks', *I3D '86 Proceedings of the 1986 workshop on Interactive 3D Graphics,* October, 89–91.

L. Pocock and J. Rosebush (2002) *The Computer Animator's Technical Handbook* (San Diego: Academic Press).

D.A. Price (2009) *The Pixar Touch: The Making of a Company* (New York: Vintage Books).

S. Prince (1996) 'True Lies: Perceptual Realism, Digital Images, and Film Theory', *Film Quarterly*, 49, no. 3 (Spring), 27–37.

DOI: 10.1057/9781137448859.0008

S. Prince (2012) *Digital Visual Effects in Cinema: The Seduction of Reality* (New Brunswick: Rutgers University Press).

L. Purse (2009) 'Gestures and Postures of Mastery: CGI and Contemporary Action Cinema's Expressive Tendencies' in S. Balcerzak and J. Sperb (eds) *Cinephilia in the Age of Digital Reproduction: Film, Pleasure and Digital Culture* (London: Wallflower) pp. 214–234.

L. Purse (2013) *Digital Imaging in Popular Cinema* (Edinburgh: Edinburgh University Press).

B. Rieder, and T. Röhle (2012) 'Digital Methods: Five Challenges' in D.M Berry (ed.) *Understanding Digital Humanities* (Basingstoke: Palgrave Macmillan) pp. 67–84.

B. Robertson (1995a) 'Untamed Animation', *Computer Graphics World*, 18, no. 4, 24–35.

B. Robertson (1995b) 'Toy Story: A Triumph of Animation', *Computer Graphics World*, 18, no. 8, 28–40, date accessed 13 July 2014.

D.N. Rodowick (2007) *The Virtual Life of Film* (London: Routledge).

D. Ryan (2011) *History of Computer Graphics* (Bloomington: AuthorHouse).

K. Salen and E. Zimmerman (2004) *Rules of Play: Game Design Fundamentals* (Cambridge, MA: The MIT Press).

W. Schaffer (2004) 'The Importance of Being Plastic: The Feel of Pixar', *Animation Journal*, 12, 72–95.

M. Seymour (2014) The art of deep compositing | fxguide, http://www.fxguide.com/featured/the-art-of-deep-compositing/, date accessed 14 August 2014.

M. Seymour (2012) MPC is berry nice to Ribena | fxguide, http://www.fxguide.com/quicktakes/mpc-is-berry-nice-to-ribena/, date accessed 15 August 2014.

M. Sims (1998) Alias Wavefront's Maya: a new generation 3-D program adds desirable tools, http://www.creativeplanetnetwork.com/news/news-articles/aliaswavefronts-maya-new-generation-3-d-program-adds-desirable-tools/381241, date accessed 21 July 2014.

T. Sito (2013) *Moving Innovation: A History of Computer Animation* (Cambridge, MA: The MIT Press).

B. Snider (1995) The Toy Story Story, http://archive.wired.com/wired/archive/3.12/toy.story.html, date accessed 1 July 2014.

M. Stahl (2005) 'Non-proprietary Authorship and the Uses of Autonomy: Artistic Labor in American Film Animation,

DOI: 10.1057/9781137448859.0008

1900–2004', *Labor: Studies in Working-Class History of the Americas*, 2, no. 4, 87–105.

G. Stern (1983) 'BBOP: A Program for Three-Dimensional Animation', *Nicograph Proceedings 1983*, 403–404.

D. Sturman (1998) 'The State of Computer Animation', *SIGGRAPH Computer Graphics Newsletter*, 32, no. 1., http://old.siggraph.org/publications/newsletter/v32n1/contributions/sturman.html, date accessed 16 April 2014.

I. Sutherland (1963) 'Sketch Pad: A Man-Machine Graphical Communication System', *AFIPS: Conference Proceedings*, 23, 329–346.

J.P. Telotte (2010) *Animating Space: From Mickey to Wall-E* (Lexington: The University of Kentucky Press).

N. Thrift (2003) 'Closer to the Machine? Intelligent Environments, New Forms of Possession and the Rise of the Supertoy', *Cultural Geographies*, 10, no. 4, 389–407.

N. Thrift (2007) *Non-Representational Theory: Space, Politics, Affect* (London: Routledge).

S. Turkle (2009) *Simulation and Its Discontents* (Cambridge, MA: The MIT Press).

J. Turnock (2012) 'The ILM Version: Recent Digital Effects and the Aesthetics of 1970s Cinematography', *Film History*, 24, no. 1, 158–168.

N. Wardrip-Fruin (2009) *Expressive Processing: Digital Fictions, Computer Games and Software Studies* (Cambridge, MA: The MIT Press).

N. Wardrip-Fruin (2011) 'Digital Media Archaeology: Interpreting Computational Processes' in E. Huhtamo and J. Parikka (eds) *Media Archaeologies: Approaches, Applications, and Implications* (Berkeley: University of California Press) pp. 302–322.

S. Whatmore (2006) 'Materialist Returns: Practising Cultural Geography In and For a More-Than-Human World', *Cultural Geographies*, 13, no. 4, 600–609.

K. Whissel (2014) *Spectacular Digital Effects: CGI and Contemporary Cinema* (Durham: Duke University Press).

H. Whitaker, J. Halas and T. Sito (2009) *Timing for Animation*, 2nd edn (Amsterdam: Focal Press).

J. Whitney (1975) Computational periodics, http://www.atariarchives.org/artist/sec23.php, date accessed 26 April 2014.

J. Wilhelms (1991) 'Dynamic Experiences' in N. Badler, B. Barksy and D. Zeltzer (eds) *Making Them Move: Mechanics, Control, and Animation of Articulated Figures* (San Mateo: Morgan Kaufmann Publishers) pp. 265–279.

DOI: 10.1057/9781137448859.0008

M.R. Wilkins, C. Kazmier and S. Osterburg (2005) *MEL Scripting for Maya Animators* 2nd edn (Amsterdam: Elsevier/Morgan Kaufmann).

M.J.P. Wolf (2012) *Building Imaginary Worlds: The Theory and History of Subcreation* (New York: Routledge).

A. Wood (2007) 'Pixel Visions: Digital Intermediates and Micromanipulations of the Image', *Film Criticism*, XXXII, no. 1, 72–94.

D. Zeltzer (1982a) 'Motor Control Techniques for Figure Animation', *IEEE Computer Graphics and Applications*, 2, no. 9, 53–59.

D. Zeltzer (1982b) 'Representation of Complex Animated Figures', *Proceedings of Graphics Interface 82*, 205–211.

Moving images cited

A Computer Generated Hand (1972, Edwin Catmull, USA) (short animation)

Angry Birds (2009–, Rovio Entertainment, Finland) (game)

BADLAND (2013, Frogmind, Finland) (game)

Biophilia (2011, Bjork, UK) (album with animations)

Cluster Movie (2005, Max Planck Institute for Astrophysics, Germany) (data visualization)

Contact (1997, Robert Zemeckis, USA) (live-action film)

Coraline (2009, Henry Selick, USA) (animated feature)

Cosmos: A Space Time Odyssey (2014, Cosmos Studios, USA) (TV programme)

Crysis series (2007– , Crytek, Germany) (game)

Dawn of the Planet of the Apes (2014, Matt Reeves, USA) (live-action film)

Despicable Me (2010, Pierre Coffin and Chris Renaud, USA) (animated feature)

Despicable Me 2 (2013, Pierre Coffin and Chris Renaud, USA) (animated feature)

Frozen (2013, Chris Buck and Jennifer Lee, USA) (animated feature)

Gertie the Dinosaur (1914, Winsor McCay, USA) (short animation)

Grand Theft Auto V (2013, Rockstar Games, USA) (game)

Gravity (2013, Alfonso Cuaron, UK) (live-action film)

Grounded: The Making of The Last of Us (2013, Area 5/Naughty Dog, USA) (live-action film)

DOI: 10.1057/9781137448859.0008

Guardians of the Galaxy (2014, James Gunn, USA) (live-action film)

Hollow Bjork (2011, Drew Berry, UK) (short animation)

How the Universe Works (2010, Discovery Channel, USA) (TV programme)

How to Train Your Dragon (2010, Dean DeBlois and Chris Sanders, USA) (animated feature)

How to Train Your Dragon 2 (2014, Dean DeBlois, USA) (animated feature)

Hugo (2011, Martin Scorsese, USA) (live-action film)

Hummingbird (1967, Charles Csuri, USA) (short animation)

Iron Man 3 (2013, Shane Black, USA) (live-action film)

Journey (2012, thatgamecompany, USA) (game)

Kung Fu Panda (2008, Mark Osborne and John Stevenson, USA) (animated feature)

L'Odyssee de Cartier (2012, Bruno Aveillan, France) (advert)

Le Faim (The Hunger) (1974, Peter Foldes, Canada) (short animation)

Marvel's The Avengers Assemble (2012, Joss Whedon, USA) (live-action film)

Matrix I (1971, John Whitney, USA) (short animation)

Matrix II (1971, John Whitney, USA) (short animation)

Metadata (1971, Peter Foldes, Canada) (short animation)

Millennium Simulation (2005, Max Planck Institute for Astrophysics, Germany) (data visualization)

Molecular Visualizations of DNA (2003, Drew Berry, Australia) (data visualization)

Monsters University (2013, Dan Scanlan, USA) (animated feature)

Monument Valley (2014, Ustwo, UK) (game)

Oblivion (2013, Joseph Kosinki, USA) (live-action film)

Osaka One, Two, Three (1970, John Whitney, USA) (short animation)

ParaNorman (2012, Chris Butler and Sam Fell, USA) (animated feature)

Permutations (1968, John Whitney, USA) (short animation)

Poem Field (1966–1969, Stan VanDerBeek, USA) (short animation)

Prometheus (2012, Ridley Scott, USA) (live-action film)

Rango (2011, Gore Verbinski, USA) (animated feature)

Rez (2001, Dreamcast, Japan) (game)

Ribena Berries (2011–, Moving Picture Company, UK) (advert)

Seeing Inside a Storm (2014, GST, Inc. – NASA/Goddard Science Visualization Studio, USA) (data visualization)

Star Trek franchise (1979–, various, USA) (live-action film)

DOI: 10.1057/9781137448859.0008

Star Wars IV: A New Hope (1977, George Lucas, USA) (live-action film)

The Hobbit: The Desolation of Smaug (2013, Peter Jackson, NZ) (live-action film)

The Last of Us (2013, Naughty Dog, USA) (game)

The Lego Movie (2014, Phil Lord and Christopher Miller, USA) (animated feature)

The Matrix (1999, The Wachowski Brothers, USA) (live-action film)

The Pony, (2013, Moving Picture Company, UK) (advert)

The Simpsons: Treehouse of Horror VI (1995, Matt Groening, USA) (TV programme)

The Writing's on the Wall (2014, OK Go, UK) (music video)

Thor: The Dark World (2013, Alan Taylor, USA) (live-action film)

Toy Story (1995, John Lasseter, USA) (animated feature)

Toy Story 3 (2010, Lee Unkrich, USA) (animated feature)

Tron (1982, Steven Lisberger, USA) (live-action film)

Up (2009, Pete Doctor and Bob Peterson, USA) (animated feature)

very.co.uk: Definitive Collection (St. Luke's Agency, UK) (advert)

Walking with Dinosaurs (1999, BBC, UK) (TV programme)

Wall-E (2008, Andrew Stanton, USA) (animated feature)

World of Warcraft: Mists of Pandaria (2012, Blizzard Entertainment, USA) (trailer)

Wreck-It Ralph (2012, Rich Moore, USA) (animated feature)

DOI: 10.1057/9781137448859.0008

Index

DOI: 10.1057/9781137448859.0009

DOI: 10.1057/9781137448859.0009

DOI: 10.1057/9781137448859.0009

DOI: 10.1057/9781137448859.0009

DOI: 10.1057/9781137448859.0009

DOI: 10.1057/9781137448859.0009